Henry S. Carhart, Horatio N. Chute

The Elements of Physics

Henry S. Carhart, Horatio N. Chute

The Elements of Physics

ISBN/EAN: 9783743318113

Manufactured in Europe, USA, Canada, Australia, Japa

Cover: Foto ©berggeist007 / pixelio.de

Manufactured and distributed by brebook publishing software (www.brebook.com)

Henry S. Carhart, Horatio N. Chute

The Elements of Physics

THE
ELEMENTS OF PHYSICS

BY

HENRY S. CARHART, LL.D.
PROFESSOR OF PHYSICS IN THE UNIVERSITY OF MICHIGAN

AND

HORATIO N. CHUTE, M.S.
INSTRUCTOR IN PHYSICS IN THE ANN ARBOR HIGH SCHOOL

REVISED EDITION

Boston
ALLYN AND BACON
AND CHICAGO

Copyright, 1892, by
HENRY S. CARHART AND HORATIO N. CHUTE.

Copyright, 1897, by
HENRY S. CARHART AND HORATIO N. CHUTE.

Norwood Press:
Berwick & Smith, Norwood, Mass., U.S.A.

PREFACE.

DURING the past decade the teaching of Physics in high schools and universities has undergone radical revision. The time-honored recitation method has gone out and the laboratory method has come in. As a natural reaction from the old *régime*, in which the teacher did everything, including the thinking, came the method of original discovery; the text-book was discarded and the pupil was set to rediscovering the laws of Physics. Time has shown the fallacy of such a method, and the successful teacher, while retaining all that is good in the new method, has already discovered the necessity of a clearly formulated, well digested statement of facts, a scientific confession of faith, in which the learner is to be thoroughly grounded before essaying to explore for himself. The maxim, "That only is knowledge which the pupil has reached as the result of experiment," has been found to have its limitations. With no previous instruction, the young student comes to the work without any ideas touching what he is expected to see, with entire ignorance of methods of experiment, and without skill in manipulation. He has no training in drawing conclusions from his own experiments. He is not a skilled investigator, and will be apt to discover little beyond his own ignorance, a result, it must be confessed, not entirely without value. Before the pupil is in any degree fit to investigate a subject experimentally, he must have a clearly defined idea of what he is doing, an outfit of principles and data to guide him, and a good degree of skill in conducting an investigation.

A few years ago it seemed necessary to urge upon teachers the adoption of laboratory methods to illustrate the text-book; in not a few instances it would now seem almost necessary to urge the use of a text-book to render intelligible the chaotic work of the laboratory.

With this view of the situation, the authors have prepared this book for the class-room, as distinguished from a manual for the laboratory. The experiments described are, for the most part, designed to illustrate principles. They are, in other words, qualitative rather than quantitative in character. The attempt to combine both characters in one book has not, in our judgment, proved a success in the past.

In the preparation of this book the aim has been to formulate clear statements of laws and principles; to illustrate them amply, both by simple experiments and by appropriate problems; and to observe a logical order and sequence of topics, so that the pupil may pass from subject to subject with the aid and momentum of what he has already acquired.

The arrangement is such that a convenient division of the book either into two or three parts, corresponding to two or three terms in the school year, may be easily made.

Every experiment, definition, and statement of principle is the result of practical experience in teaching classes of various grades. Many of the illustrations are new, having been made from original drawings or from photographs of the actual apparatus set up for the purpose.

We should recommend teachers to supplement the class-work in this book with practical work in the laboratory. It is not necessary that the pupil should traverse the entire subject of Physics before taking up laboratory practice, but he should be kept in his class-work well ahead of the subjects forming the basis of his laboratory experiments.

An entire year should be devoted to the subject, and great advantage will be found in delaying its introduction till the student has acquired some knowledge of algebra and geometry.

H. S. C.
H. N. C.

ANN ARBOR, MICHIGAN.

CONTENTS.

	PAGE
Introduction	1

CHAPTER I.

The Properties of Matter	6

CHAPTER II.

Mechanics of Solids		25
I.	Composition of Velocities	25
II.	Newton's Laws of Motion	32
III.	Work and Energy	44
IV.	Gravitation. — Centre of Mass. — Stability	50
V.	Curvilinear Motion	57
VI.	Accelerated Motion. — Falling Bodies	59
VII.	The Pendulum	63
VIII.	Simple Machines	70

CHAPTER III.

Mechanics of Fluids		83
I.	Nature of Fluids	83
II.	Transmission of Pressure	84
III.	Pressure due to Gravity	87
IV.	Pressure on the Bottom of a Vessel	89
V.	Equilibrium in Fluids	92
VI.	The Barometer	94
VII.	The Air-Pump and Condenser	98
VIII.	Law of Boyle	102
IX.	The Siphon and the Water-Pump	104

CONTENTS.

Mechanics of Fluids, *continued.*

		PAGE
X.	The Principle of Archimedes	110
XI.	Density and Specific Gravity	113

CHAPTER IV.

Heat 119

I.	Nature of Heat	119
II.	The Thermometer	121
III.	Sources of Heat	126
IV.	Distribution of Heat	128
V.	Effects of Heat	140
VI.	Calorimetry	153
VII.	Heat and Work	157

CHAPTER V.

Magnetism and Electricity 161

I.	Magnets. — Polarity. — Induction	161
II.	Nature of Magnetism	166
III.	The Magnetic Field	169
IV.	Terrestrial Magnetism	172
V.	Static Electricity	174
VI.	Induction	180
VII.	Electrical Distribution	181
VIII.	Electrical Capacity	184
IX.	Electrical Machines	189
X.	Experiments with Electrical Machines	195
XI.	Atmospheric Electricity	198
XII.	Current Electricity	200
XIII.	Effects of Electric Currents	211
	a. Heating Effects	211
	b. Chemical Effects	212
	c. Magnetic Effects	216
XIV.	Electrical Quantities	221
	a. Ohm's Law	221
	b. Instruments for Measurement	228
XV.	Current Induction	235
XVI.	The Dynamo	244
XVII.	The Electric Light	250

CONTENTS. ix.

	PAGE
Magnetism and Electricity, *continued.*	
XVIII. Electric Motor and Telegraph	253
XIX. The Telephone and Microphone	256

CHAPTER VI.

Sound	260
I. Wave-Motion	260
II. Sources of Sound	266
III. Transmission of Sound	268
IV. Reflection and Refraction of Sound	270
V. Intensity and Loudness	273
VI. Velocity of Sound	277
VII. Interference	279
VIII. Sympathetic Vibrations	282
IX. Pitch	284
X. Vibrations of Strings	289
XI. Overtones and Harmonic Partials	292
XII. Vibration of Air in Tubes	294
XIII. Quality or Timbre	297
XIV. Harmony and Discord	299
XV. Vibrating Rods, Plates, and Bells	300
XVI. Graphic and Optical Study of Sound	302

CHAPTER VII.

Light	306
I. Its Nature	306
II. The Propagation of Light.—Its Velocity	307
III. Photometry	312
IV. Reflection of Light	314
V. Refraction of Light	329
VI. Lenses	339
VII. Dispersion	350
VIII. Color	356
IX. Spectrum Analysis	360
X. Interference of Light	364
XI. Optical Instruments	367

Appendix.

	PAGE
Diagrams of Metric Measure	375
Tables for converting to and from Metric System	376
Tables of Constants for Force, Work, and Activity	380
Table of Densities	381

Index 383

ELEMENTS OF PHYSICS.

INTRODUCTION.

1. **Matter** may be defined as that which occupies space. Its existence is made known to us through the senses. *Substances* are the different kinds of matter, as iron, water, air.

2. **Its Structure.** — Many facts, a few of which will be considered later, indicate that matter is not continuous, and that any portion of it is a group of exceedingly small parts called *Atoms*, no two of which are in actual contact. These atoms are kept from falling apart by the action of certain forces or agents which bind them together in groups, called *Molecules*. The nature of an atom is such that no division of it occurs during any physical or chemical change; but, while the molecule is unaffected in any physical change, it is broken up or has the mode of grouping of its constituent atoms changed when it is subjected to a chemical change.

These molecules are believed to be in ceaseless motion, subject to the constraining action of certain forces, called *Molecular Forces*. These motions are supposed to account for many of the phenomena observed in the world of matter.

Although molecules are so small that microscopes of the highest power cannot detect them, yet they are not beyond the limits of calculation. Lord Kelvin has estimated that the average distance between the centres of two adjacent molecules of a liquid or a solid is less than the $\frac{1}{5,000,000}$th, and greater than $\frac{1}{1,000,000,000}$th of a centimetre.[1]

3. Substances Classified. — If we place an iron nail between the jaws of a blacksmith's vise, and subject it to pressure, we shall find, on removing it, that very little effect has been produced; the sides which were free from pressure may have bulged out somewhat, and the nail may have been slightly flattened. But if we try to substitute water or air for the iron, we shall find that each escapes laterally so readily that we cannot subject either of them to pressure unless we adopt some means of retaining it on all sides. Hence it appears that we may make two classes of substances, — *Solids*, which can sustain pressure without being supported laterally, and *Fluids*, which cannot do so.

Again, if a small quantity of water be put into an empty vessel, it will partly fill it, and assume the shape of that part of the vessel occupied by it. Such fluids are called *Liquids*. If, however, a small quantity of air be put into a closed and empty vessel, it will expand and fill the vessel, irrespective of its size. Such fluids are called *Gases*.

All simple and many compound substances, as water, can be made to pass successively through all three states by suitable adjustments of temperature and pressure.

---

[1] Thomson's Constitution of Matter, Vol. I., page 217.

4. **Energy.**—It is a matter of daily observation that a body in motion can impart that motion to a second body; as, for example, when a swinging bat strikes a ball. It is equally true that a body can set a second one in motion by virtue of the position which it occupies, as in the case of the weight of a clock when it is wound up. This capability of a body of producing motion in a second one is known as *Energy.*

5. **Physics** is that branch of science which treats of *matter* and *energy*. Its investigations are restricted to the examination of those changes in which the composition of the molecule is not affected. It differs from *Chemistry* in that it does not deal with intramolecular changes nor concern itself with the identification of substances.

6. **Physical Phenomenon, Law, Theory.**—Any change observed in matter is a *Phenomenon*. If unattended by any alteration of the chemical constitution of the substance, the change is a *Physical* one; otherwise, it is *Chemical*. The fall of an apple or of a rain-drop, the freezing of water, the melting of a metal, the expansion of a metallic rod when heated, the attraction of an iron nail by a magnet, are examples of *Physical Phenomena*.

A *Physical Law* is the expression of the constant relation existing between dependent phenomena. For example, it is found on trial that if a quantity of any true gas, as air, be subjected to pressure, the volume will diminish as the pressure increases. This relation is known as the *Law of the Compressibility of Gases*.

A *Hypothesis* is a supposition advanced to explain phenomena. The probability of its truth is the greater, the more varied are the phenomena explained. When the amount of evidence in support of a hypothesis is quite large, it is raised to the rank of a *Theory;* and finally to that of a *Law*, when its truth is fully established.

7. Physical Agents. — In attempting to ascertain the causes of phenomena, it seems necessary to assume the existence of certain *Agents* or *Forces* acting on matter. Examples of such agents are seen in the various forms of attraction of matter for matter. Since nothing is known of the intimate nature of these agents, all attempts at defining them must be restricted to narrations of effects.

8. Force and Forces Classified. — *Force* may be defined as that which changes, or tends to change, the motion of a body by altering either its direction or its magnitude. Forces may be classified as *Molar Forces*, those acting between masses separated by sensible distances, as gravitation; *Molecular Forces*, those acting between molecules separated by insensible distances, as cohesion; and *Atomic Forces*, those acting between the atoms composing the molecule, as chemism. Again, forces may be classified as *Attractive Forces*, as the action between magnets and soft iron; and *Repellent Forces*, as the action between two pieces of sealing-wax after both have been rubbed briskly with flannel. Finally, forces may be classified as *Impulsive Forces*, where the duration of their action is exceedingly short, as when a nail is struck with a hammer; and *Continuous Forces*, when their action extends over a finite time, as gravity. If the intensity of a continuous force varies during successive intervals of time, it is classed as a *Variable Force;* for example, the attraction between a magnet and a piece of iron becomes less on increasing the distance between them. If the intensity of a force does not change, it is said to be a *Constant Force;* for example, the attraction of the earth for a body on its surface changes so little when the body is removed a short distance from the surface that it may be considered constant within short ranges without sensible error.

9. An Experiment consists in producing a phenomenon under conditions determined and regulated by the operator. By making measurable changes in these conditions and noting the results, he is able to determine the conditions under which the phenomenon takes place, and the class to which it belongs; or to discover the general law which embraces it. The basis of all experimentation is a belief in the *Constancy of Nature*, as enunciated in the postulate that " *under the same physical conditions the same physical results will always be produced irrespective altogether of time or place.*"

CHAPTER I.

THE PROPERTIES OF MATTER.

10. The Properties of Matter are the different ways in which it presents itself to our senses; as *occupying space, requiring force to move it*, etc.

11. Physical and Chemical Properties.—Experiment.— Hold a piece of platinum wire in the flame of a Bunsen lamp. It becomes red-hot, then glows and emits light. Remove it, and it soon resumes the same appearance that it had before it was heated. In like manner, hold a piece of magnesium wire in the flame. It, too, glows, but soon begins to burn, leaving a residuum in the form of a white powder.

Two classes of changes are illustrated by this experiment. In the case of the platinum there is no loss of identity, the molecule remaining intact; in the case of the magnesium, the composition of the molecule is changed by the introduction of oxygen into its structure. Properties revealed by changes in bodies where the molecule suffers no change of composition are called *Physical Properties*. Those revealed by bodies during changes in the composition of the molecule are *Chemical Properties*.

12. Extension, as applied to matter, means that it occupies space. Every body, or piece of matter, has *length*, *breadth*, and *thickness*, and these are known as its *Dimensions*.

13. Measurement of Extension. — The basis of all measurement is some linear unit, arbitrarily chosen, from which are derived the units of area and of volume. There are two systems of units in general use, the *English* and the *Metric*. Under the former, the *Yard* is the standard of length. For the United States this is defined as the distance between the 27th and the 63d divisions of a brass bar deposited in the office of Weights and Measures at Washington, provided the temperature is $16\frac{2}{3}°$ C. One thirty-sixth of this constitutes the *Inch*, and one-third the *Foot*. Under the metric system, the *Metre* is the standard of length, and may be defined as the distance between two lines on a platinum-iridium bar at $0°$ C., also kept at Washington. This system was legalized in the United States in 1866. In its multiples and submultiples a decimal system is adopted; the names of these derived units being formed by the use of the Greek prefixes *deca*, *hecto*, and *kilo*, signifying 10, 100, and 1000; and the Latin prefixes *deci*, *centi*, and *milli*, signifying 10th, 100th, and 1000th.

The units of surface are squares whose sides are some one of the units of length, as the square inch, square foot, square metre, etc. The units of volume are cubes whose edges are some one of the linear units, as the cubic inch, cubic foot, cubic centimetre, etc. The United States *gallon*, 231 cubic inches, and the *litre*, 1000 cubic centimetres, are units of volume, but have no fixed form.

The legal value of the metre is 39.37 inches. When great accuracy is not required, it may suffice to regard the metre as 40 inches, the decimetre as 4 inches, the centimetre as $\frac{2}{5}$ of an inch, the kilometre as $\frac{5}{8}$ of a mile, and the litre as $2\frac{1}{9}$ pints.

In measuring any object various devices and methods

are in use, the selection being determined by the nature of the problem and the degree of accuracy required.

14. Weight and Mass. — The force of gravitation is universal. It applies to all matter at all distances. It explains the motions of the heavenly bodies and the tendency of all bodies to fall to the earth. It is this force which gives rise to weight. The *Weight* of a body is the measure of the attraction of the earth for it; it varies with the quantity of matter in it, and with the position of the body with respect to the earth's surface. Hence, the weight of a body is not a constant quantity. The *Mass* of a body is the quantity of matter it contains; unlike weight, it is wholly independent of the attraction of any other body. Hence, the mass of a ball of lead would remain the same if it were transported to the centre of the earth, to the north pole, or to the moon; whereas its weight would be different for each position.

15. Their Measurement. — The unit of mass adopted in this country and in England is the quantity of matter in the *Avoirdupois Pound*. It is equivalent to 7,000 grains. Its chief multiples and submultiples are *tons*, *ounces*, and *drams*. In the metric system the *Gramme* is the unit, and may be defined as the quantity of matter in one cubic centimetre of distilled water at $4°$ C. It is equivalent to 15.432 grains. It is multiplied and subdivided decimally, and the names given to these multiples and submultiples are formed by combining the word *gramme* with the prefixes used in linear measure (13).

The units of weight are the attractions of the earth for these units of mass, and receive the same names. What is commonly termed the *weight* of a body, obtained by counterpoising it with standard units of mass, is its *mass*.

The weight might be determined by some such device as the common draw-scale (Fig. 18), where the attraction is measured by the elongation of a spring. Even then it must not be overlooked that a comparison of weights thus obtained, when under exactly similar conditions, may be substituted for a comparison of masses, since at any one place *the weight varies as the mass.*

16. **Impenetrability** means that two masses of matter cannot occupy the same space at the same time. The familiar method of measuring the volume of an irregular solid by measuring that of the liquid displaced when the solid is submerged is based on this property.

According to the accepted theory of the constitution of matter (2), a body does not fill completely the space which it appears to occupy, that is, there are spaces or *pores* between its parts. Hence, there is suggested the possibility of an *interpenetration* between two masses of matter when brought together, the molecules of the one fitting into the spaces between the molecules of the other, and the resulting volume not being equal to the sum of the two volumes. An instance of this is seen in the following experiment:

In a long test-tube pour 30 ccm. of water. To this add carefully 20 ccm. of strong alcohol, tipping the tube so that the liquid flows down its walls. Mark the position of the surface, and then mix the liquids thoroughly by shaking. A large shrinkage in volume is the result.

Hence, it appears that impenetrability, though always true of molecules, is not necessarily a property of masses.

17. **Inertia** is that property manifested by matter in its persistence in whatever state of rest or of motion it may happen to have, and in its resistance to any attempt to

change that state. If a moving body be stopped, its arrest is always traceable to some influence outside of that body; if a body left at rest be afterwards found in motion, such motion has been imparted to it by some other body. There are many familiar phenomena which are explained by this principle; for example, the onward motion of a rider when his horse suddenly stops; the displacement of loose articles on a rolling ship; the oscillation of water in a pail when carried; the mud flying from a rapidly revolving carriage wheel; the dust from a carpet when beaten; snow falling from the boot when struck against the door-step. The inertia of a stream of water is seen in the violent jar to the pipe on suddenly closing the faucet. An instance of inertia is furnished in the persistence with which a rotating object, as a spinning top, maintains its plane of rotation. It is because of inertia that the suspended hammock on ship-board does not fully partake of the motion of the ship.

Fig. 1.

By inertia may be explained the following experiment: Suspend by a string a heavy weight, A (Fig. 1), and from its under side by a similar string a small bar, B. If we pull *steadily* downward on B, the string will break above A. The tension in the upper string is greater than in the lower one, because it has to support A and resist the pull applied at B. If, however, we pull downward *suddenly* on *B*, the string will break below *A*. On account of the inertia of the heavy weight, the lower string breaks before the sudden pull reaches the upper string.

18. Indestructibility. — **Exp.** — Fill a test-tube, say 2 cm. in diameter and 15 cm. long, with water; cover the mouth with the thumb, and invert the tube in a small beaker partly filled with water (Fig. 2). There must be no air in the tube after it is inverted. Slip over the tube a short piece of large glass tubing — a piece of lamp chimney will answer — to support the test-tube in a vertical position. Place a small piece of zinc, say, 50 milligrammes, beneath the mouth of the tube. Pour into a small beaker a few cubic centimetres of strong sulphuric acid. Place the vessel supporting the tube, and also the beaker of acid, on the pan of a good balance, and counterpoise them. Now pour the acid into the vessel supporting the tube, replacing the beaker in the scale-pan. The acid will act on the zinc, and hydrogen gas will collect in the top of the tube. If from time to time the beam of the balance be released, it will be found that the beaker, with its contents, loses nothing in weight during the disappearance of the zinc; that is, *there is no loss of matter during these changes.*[1]

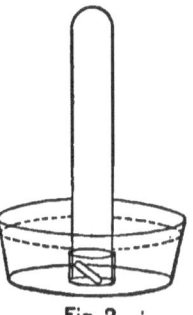

Fig. 2.

Indestructibility implies that matter cannot be destroyed. All experience testifies that we can neither create nor destroy matter; we can at most but change its form and its combinations. The science of chemistry is based on this *Constancy or Conservation of Mass.*

19. Elasticity. — **Exp.** — Apply pressure to a rubber ball. Stretch a common rubber band. Bend a thin strip of steel. Twist a piece of cotton cord. In each case either the form or the volume has been changed, and unless the distorting force is maintained these changes disappear when the body resumes its initial condition.

---

[1] A delicate balance will show an apparent loss, because the disengagement of hydrogen causes an increased displacement of air.

Elasticity is that property exhibited by matter when a continued application of force is necessary to maintain any change of form or volume which has been produced by a stress. It is called *Elasticity of Form* when the form is restored on the removal of the distorting force, and *Elasticity of Volume*[1] when the initial volume is recovered. Evidently fluids do not possess elasticity of form, but all matter possesses elasticity of volume, some kinds in a greater degree than others. Fluids regain exactly their original volume, without reference to the duration or the amount of the distortion. In the case of solids the restitution is not complete, if the distortion is beyond a certain limit, different for different substances.

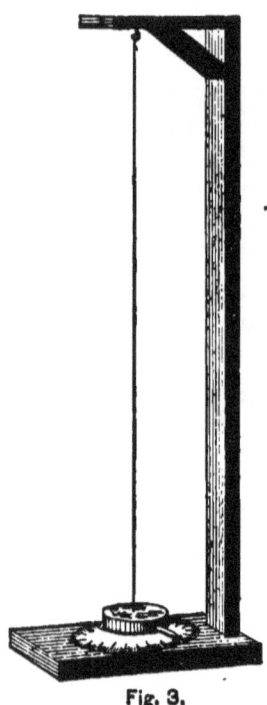

Fig. 3.

Experiments show that elasticity of form or of volume may be developed either by *pressure*, by *stretching*, by *bending*, or by *twisting*. The bounding ball and the common pop-gun are illustrations of the first; rubber cords are familiar examples of the second; bows and springs are instances of the third; and the torsional pendulum (Fig. 3) and the stretched spiral spring illustrate the fourth.

---

[1] *The Coefficient of Elasticity of Volume* is the ratio of the increase of pressure per unit of area of surface to the compression produced per unit of volume. Let p be the increase of pressure and s the area of the surface, then $\frac{p}{s}$ is the pressure per unit of area. Let V be the original volume and v the resulting volume; then $V-v$ is the magnitude of the compression, and $\frac{V-v}{V}$ is the compression per unit of volume. Hence $\frac{p}{s} \div \frac{V-v}{V} = \frac{pV}{s(V-v)}$ is the coefficient of elasticity of volume.

20. Cohesion. — Exp. — Suspend beneath one of the scale-pans of a beam balance a perfectly clean glass disk by means of threads cemented at three equidistant points (Fig. 4). After counterpoising the disk, place below it a vessel of water and adjust the apparatus so that the disk touches its surface when the beam is horizontal. Now add weights to the pan till a separation of the disk from the water is effected. An examination of the under surface of the disk will show that instead of pulling the plate away from the water, we have pulled off a film of water.

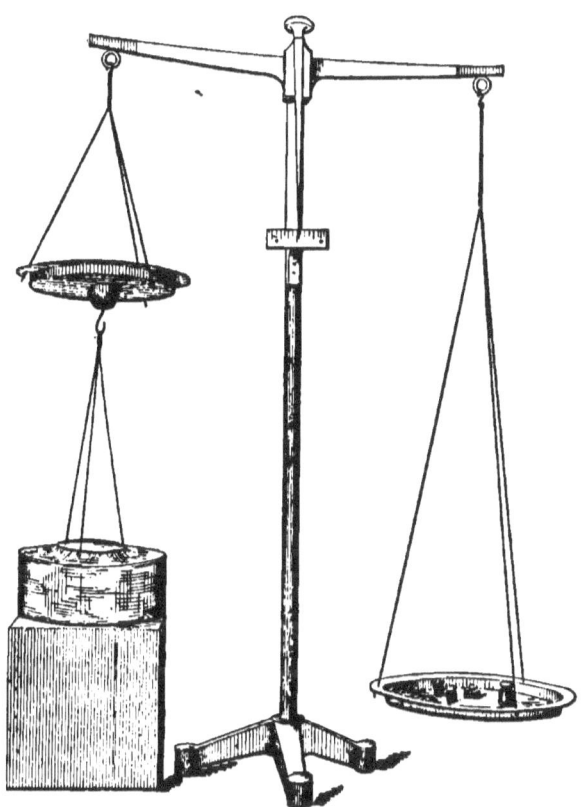

Fig. 4.

The experiment makes it clear that the parts of the water are held together by a force of considerable intensity, and that this force has to be overcome before a separation can be effected. To that force which binds together the molecules of a body, the name *Cohesion* or *Adhesion* is applied according as the molecules are like or unlike in character.[1]

---

[1] It is not practicable to maintain this distinction at all times; as in the case of a solution of a substance, what name shall be applied to that agent which holds two parts of the liquid together? It is doubtful if clearness is gained by such a refinement of terms, and furthermore it is very doubtful whether there is any difference between them.

This force exhibits itself more noticeably in solids than in liquids; but in gases it seems to be nearly, if not entirely, wanting. Generally the attraction of liquids for solids is greater than that of the liquid for itself, as seen in the above experiment. This is always the case when the liquid *wets* the solid. Sometimes, however, the solid can be pulled away from the liquid, as in the case of glass and mercury, and this shows that the attraction of mercury for mercury exceeds its attraction for glass.

21. **Affected by Distance.**—Exp.—Press firmly together two lead disks, giving them a slight twisting motion. The surfaces in contact must be flat and bright. They will adhere quite firmly together.

Two lead bullets, cut so as to present clean flat surfaces, may be used instead of the disks.

This experiment indicates that the molecular forces act powerfully through exceedingly small and insensible distances; and that they act either not at all or very feebly through distances observable by the eye. They seem to diminish rapidly in intensity as the distances between the molecules are increased, as seen in the effect of temperature on the strength of materials.

Queries:—Explain the welding of iron. Why is it that powdered substances can be solidified by great pressure? Why do drops of different liquids vary in size?

22. **Crystallization.**—Exp.—Dissolve 100 grammes of alum in half a litre of hot water. Hang several strings in the solution and set it aside in a quiet place for several hours. At the end of that time the strings will be found covered with a number of beautiful, transparent bodies of regular and similar shapes.

Most solids, when they form slowly under circumstances that give perfect freedom of motion to the molecules, assume definite structural forms called *Crystals*. These forms are, with certain limitations, the same for the same substance, but different for different substances. Many substances, however, exhibit no plan in the grouping of the molecules, that is, they are *Amorphous*. Examples of such bodies are glass, glue, coal, etc. When crystalline substances are broken into parts, it will be found that the fracture consists in separating crystals from the faces of other crystals. This separation is most easily effected along certain definite planes, called *Planes of Cleavage*.

23. **Tenacity.**—Exp.—Cut from a sheet of manilla paper a rectangular strip, say, 25 cm. long and 5 cm. wide. Fold over each end and fasten it with glue to form a hem or loop. In each of these loops insert a stout wooden rod, somewhat longer than the width of the paper. Connect the ends of one of these rods to the hook of a spring-balance (54) by means of a stout wire bail. Fasten the other to some suitable object, as a hook in wall. Now pull steadily on the ring of the balance till the paper breaks, and observe the reading of the balance at the moment of separation. The average of several trials will give a fair idea of the strength of the paper. If a strip of paper 10 cm. wide be used, we shall find that about twice as great a pull will be required to tear it as in the first case.

Tenacity is the resistance which a substance offers to being separated into parts by pulling. Its measure is the quotient of the breaking-weight by the area of the cross-section of the substance broken. Experiments show that it varies with different substances, with the form of the body, with the temperature, and with the duration of the traction.

24. **Malleability** is that property of a body which permits it to be rolled or hammered into thin sheets. Gold furnishes the best illustration of this property. It has

been reduced to sheets so thin that more than 300,000 of them placed one upon another are but one inch in thickness. This property in some substances is much modified by temperature.

25. Ductility is that property of a body which permits it to be drawn out into wire. Malleability and ductility are closely related in character, but do not exist in the same substance to the same degree; for example, lead and tin are very malleable, but only slightly ductile. Generally, however, substances possessing great malleability are also highly ductile, as in the case of gold, platinum, and silver.

26. Hardness is the relative resistance which a body offers to scratching or abrasion by other bodies. The relative hardness of two substances is ascertained by trying which of them will scratch the other; for example, glass is harder than copper, since it scratches copper.

Many substances, if suddenly cooled after having been raised to a high temperature, acquire great hardness. A few, however, as copper and bronze, are affected in a contrary manner. The process of imparting to a body a suitable degree of hardness is called *Tempering;* that of making it as soft as possible at ordinary temperature is called *Annealing.* Both processes consist in raising the temperature of the substance and then cooling, either suddenly or slowly, according to the result desired.

27. Nature of Surface of a Liquid.—Exp.—Place a sewing-needle on the surface of a vessel of water. If carefully done, the needle will float. On close examination it will be seen that the surface of the water round the needle is depressed, the latter resting in a little hollow large enough to hold, perhaps, four such needles.

The indentation made in the surface of the liquid by the needle suggests that the surface is like a membrane or skin, supporting the needle by virtue of its cohesion or toughness.

28. In a State of Tension. — **Exp.** — Float two bits of wood on water, placing them parallel and separated by a few millimetres. Let a drop of alcohol fall on the water between them, and they will suddenly fly apart.

The surface layer of the water acts as if it were a stretched membrane, the effect of the alcohol being to weaken or cut it between the splinters and thereby permit the parts to separate, carrying the splinters with them.

Exp. — Place a thin layer of water on glass and let a small drop of colored alcohol fall on it. The weak spot made by the alcohol will cause a rupture in the film, and the water will be drawn away, leaving the alcohol surrounded by a dry area.

Exp. — Make a ring, 3 or 4 inches in diameter, of stout iron wire, with a supporting handle. Tie to this a loop of thread, so that the loop may hang near the middle of the ring (Fig. 5). If now the ring is dipped into a soap solution, and a film is obtained in it, the loop of thread will spring out into a circle when the film inside the loop is carefully broken. The tension in the film pulls the thread outward in all directions equally. By tilting the ring the circle will float about on the film.

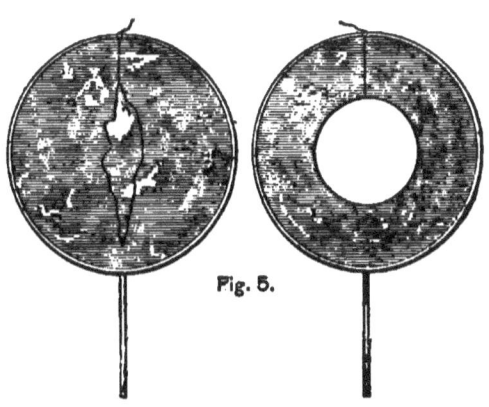

Fig. 5.

29. Surface Tension accounted for. — The molecules composing the surface of a liquid are not under the same conditions as those within the liquid, since the latter are attracted equally in all directions by the surrounding molecules; whereas those composing the surface layer are attracted downward and laterally, but not upward. The result is that this surface layer of molecules is under a tension resembling a stretched skin, and at the same time exerts a pressure on the liquid below. On lessening the cohesion of this surface film at any point, the other parts are pulled away. To this is due the movement of the splinters in the above experiment. The peculiar gyratory motions of a piece of camphor floating on clean, warm water is another interesting illustration of the same principle, the slight and unequal solubility of the camphor weakening the supporting film with the results described.

30. Form of Surface. — **Exp.** — Fill a goblet two-thirds full of water. An examination of the surface reveals the fact that the surface is level, except at the edge next the glass, where it curves upward (Fig. 6). The water adheres to the glass, and climbs up the sides of the goblet. If we fill a similar goblet with mercury, we find that the surface is convex next to the glass, as if the liquid were pulled away from it.

Fig. 6.

When the liquid wets the vessel, that is, when the attraction between it and the vessel exceeds that between the molecules of the liquid, the surface is concave; on the contrary, when the liquid does not wet the vessel, that is, when the attraction between the molecules of the liquid exceeds that between it and the vessel, the surface is convex.

THE PROPERTIES OF MATTER.

31. Capillarity.— Exp.— Support vertically several clean glass tubes of small and different diameters in a vessel of pure water, first wetting their inner surfaces to displace the air-film on the glass surface (Fig. 7). The water will be seen to rise in these tubes, highest in the one of smallest diameter and least in the one of greatest. If mercury be used, instead of water, it will be depressed in the tubes, the most in the smallest tube. On examining the surface of the liquid within the tubes, it is found to be concave when the liquid rises and convex when it is depressed.

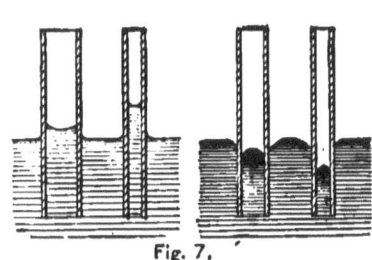
Fig. 7.

Capillarity or *Capillary Action* (*capillus*, a hair) is the name given to the phenomenon of the rising or falling of liquids in fine, hair-like tubes.

32. Laws of Capillary Action.— The following laws of capillary action have been established by experiment:

I. *Liquids ascend in tubes when they wet them, that is, when the surface is concave; and they are depressed when they do not wet them, that is, when the surface is convex.*

II. *The elevation or the depression is inversely as the diameter of the tube.*

III. *The elevation or the depression decreases as the temperature increases.*

33. Influence of the Curved Surface.— Exp.— Introduce a drop of water into the large end of a conical capillary tube, made by drawing out a larger tube after heating it in a gas flame. The surfaces in contact with the air assume a concave form (Fig. 8), the *smaller one* having the *greater* curvature, and the water moves toward the *smaller* end of the tube.

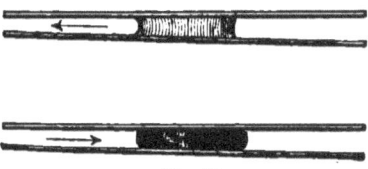
Fig. 8.

Into the small end of a similar tube put a drop of mercury. The free surfaces will be convex, the *smaller one* having the *greater* curvature, and the mercury moves toward the *larger* end of the tube.

Each free surface of the liquid exerts either traction or pressure on the liquid, according as it is concave or convex. The magnitude of this surface action increases with the curvature, as shown by the drop of liquid in the first case moving toward the smaller end of the tube, and in the second case moving away from it.

34. Capillarity Explained. — When the liquid wets the tube, the surface of the liquid within the tube is concave, and hence exerts an upward pull on the liquid, causing it to rise in the tube to a point where the weight balances this force of traction. When the liquid does not wet the tube, the surface is convex and hence exerts a downward pressure, thereby causing a depression of the liquid within the tube.

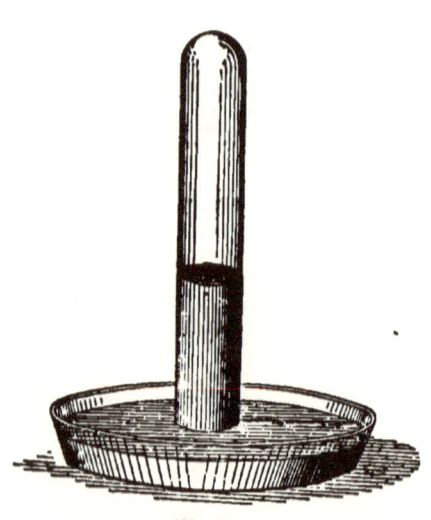

Fig. 9.

35. Capillary Phenomena. — Familiar illustrations of capillary action are numerous. The action of blotting-paper in removing an excess of ink, the passage of oil up the wick of a lamp, the absorption of water by a sponge or cloth, are among the more common instances.

For further discussion of surface tension and its relation to capillarity, the student is referred to Maxwell's Theory of Heat, Chap. XX., Tait's Properties of Matter, Chap. XII., and Daniell's Physics, Chap. XI.

36. Absorption.—**Exp.**—Fill a stout, wide test-tube with dried ammonia gas by displacement over mercury. The gas may be obtained by boiling strong ammonia-water in a flask and passing it through a tube filled with small lumps of quick-lime. Insert beneath the mouth of the test-tube a piece of recently heated charcoal, and immediately the mercury will begin to rise in the tube, owing to the absorption of the gas by the charcoal (Fig. 9).

Absorption is the penetration of any body into a porous one. Freshly burnt charcoal in a most remarkable degree possesses this power of absorbing gases. Gases which are most readily liquefied by pressure are, as a rule, absorbed by porous solids to the greatest extent. It should be noted that this absorption is greatly influenced by temperature and pressure, being decreased by an increase of temperature, and increased by an increase of pressure; and also that the presence of a gas in the pores of a solid affects the amount of another gas that the body will absorb, but not always in the same way.

The persistence with which air adheres to the surface of a solid is illustrated by the barometer tube (145). It is found necessary to boil the mercury in the tube in order to expel the air completely. It also affects capillarity, modifying the heights to which liquids rise, as shown in the increased height following the displacement of the air-film by a liquid film, on wetting the walls of the tube.

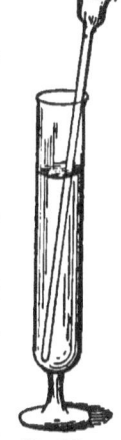

Fig. 10.

37. Diffusion.—**Exp.**—Fill a large test-tube two-thirds full of water, colored with blue litmus. Introduce a few drops of sulphuric acid into the liquid at the bottom of the tube by means of a thistle-tube (Fig. 10). A reddish color will be seen at the bottom; and, if the liquids are not disturbed for several hours, this change of color will be seen to move slowly toward the top.

Exp. — Fill two six-ounce wide-mouthed bottles with hydrogen and oxygen respectively, having previously fitted to them perforated corks through which passes a glass tube about 60 cm. long. After connecting these two bottles by this tube, set them in a vertical position, with the one containing the hydrogen uppermost. After the lapse of an hour, if we take them apart and apply a lighted taper to each jar, an explosion will follow, showing that the gases have mixed.

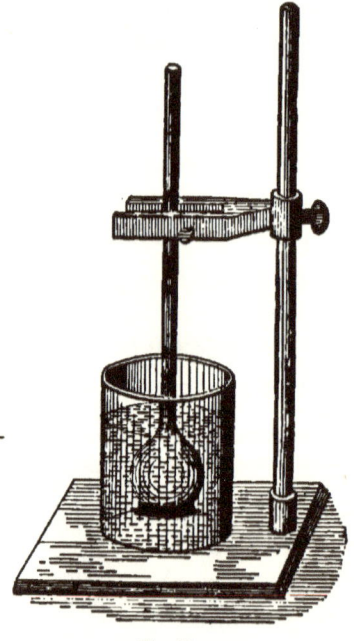

Fig. 11.

In both of these experiments we have the heavier fluid slowly moving upward through the lighter one, and that in turn moving downward into the heavier. Such an intermingling of fluids, when placed in contact, is called *Diffusion*. Such phenomena offer a striking confirmation of the theory that the molecules of fluids are in motion. Experiments show that the rate of diffusion is affected both by density, temperature, and the miscibility of the substances.

38. Osmose. — **Exp.** — Wet a piece of parchment paper and tie it across the bowl of a large thistle-tube (Fig. 11). Pour down the stem of the tube a concentrated solution of copper sulphate. Support the tube in a beaker of water, so that the liquids are at the same level on both the inside and the outside of the tube. If we watch the liquids for a short time we shall find that the one within the tube is slowly rising and the one without the tube is acquiring a bluish tint, showing that the two liquids are passing through the membrane, with the greater flow toward the denser liquid.

Exp. — Cement a small porous battery-cup to a funnel-tube. Connect it to a wide-mouthed bottle provided with a jet-tube, after the plan shown in Fig. 12. Over the porous cup invert a jar of hydrogen gas. If all the joints are gas-tight a small fountain will be caused by the pressure of the hydrogen gas on the water in the bottle. The gas evidently passes through the walls of the porous cup faster than the air passes out.

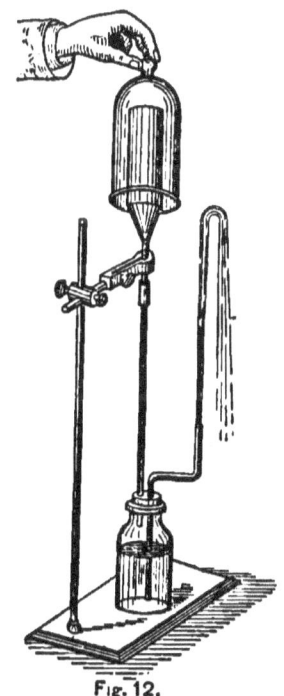

Fig. 12.

That variety of diffusion wherein the fluids pass through a porous membrane is called *Osmose;* the flow of the fluid toward the one which increases in volume being called *Endosmose*, and the contra-current *Exosmose*. The phenomenon is dependent to a large extent on the nature of the membrane, and seems to be closely related to capillarity on the one hand and to diffusion on the other.

39. **Dialysis.** — It was discovered by Dr. Graham that all kinds of salts which dissolve in water diffuse easily through porous membranes; but substances which have no crystalline structure, as gums, starches, etc., diffuse with extreme slowness.

24   ELEMENTS OF PHYSICS.

Substances of the former class are called *Crystalloids*, and those of the latter *Colloids*. Chemists often separate colloids from crystalloids by placing the mixture in a vessel having a bottom of parchment paper, the whole being suspended in a vessel of water. This process is called *Dialysis*, and the apparatus a *Dialyser*.

PRACTICAL QUESTIONS.

1. Why is the point of a pen slit?
2. Why will not writing-paper absorb ink?
3. Why is a person, leaping to the ground from a rapidly moving carriage, likely to fall in the direction in which the carriage is going?
4. In applying benzol to a grease spot on cloth, why should we begin around the edge of the spot and work up toward the centre?
5. What objection to the method of measuring medicines by drops?
6. Account for the possibility of carrying water in a dish whose bottom is a fine-meshed wire cloth.
7. Carbonic acid is considerably heavier than air. Why does it not collect at the bottom of inhabited rooms?
8. Why does water rise higher in a capillary glass-tube than any other liquid?
9. If a soap-bubble be blown at one end of a tube, it will rapidly contract, driving a current of air out of the other end of the tube. Explain.
10. State the molecular theory of the construction of matter, and point out what facts presented in the preceding pages seemingly testify to its validity.
11. What property of matter is recognized in measuring the volume of a solid by noticing the volume of liquid it displaces? When does this method fail?

CHAPTER II.

MECHANICS OF SOLIDS.

I. COMPOSITION OF VELOCITIES.

40. Mechanics is that branch of Physics which treats of force and its effects on bodies. The study of the conditions producing rest in bodies belongs to *Statics;* and of those producing motion, to *Kinetics*.

41. Motion is the continuous change in the relative position of a body with respect to some point or place of reference. The path of a moving body must be continuous, that is, without any interruptions, since the body has continuous existence and must be somewhere at every instant. All motion is relative, since there are no fixed points in space to which absolute motion can be referred. The motion of a train of cars is relative with respect to the earth's surface; that of the earth's surface is relative with respect to its centre; and the motion of its centre is relative with respect to the sun.

When the path of a moving body is a straight line, its motion is *Rectilinear*, and when curved it is *Curvilinear;* in the latter case the direction of its motion at any point of its path is that of the tangent to the curve at that point.

42. The Velocity of a moving body at any instant is the distance it would pass over in the next unit of time, if left wholly to itself. For example, the velocity of a bullet on leaving a rifle is the distance it would pass over in a second, if nothing disturbed it during that second; the velocity of a falling stone at any moment is the distance it would pass over during the following second, if the attraction of the earth and the impeding action of the air could both be withheld during that second.

When a body in motion moves over equal distances in successive equal intervals of time, its motion is *Uniform;* otherwise it is *Variable*. If the velocity of a body increases during each successive interval of time, the motion is *Accelerated;* if it decreases, it is *Retarded*. *Acceleration is the change of velocity* per unit of time. It is positive when the velocity increases, and negative when it decreases. When the acceleration is constant, the motion is *Uniformly Accelerated* or *Retarded*, according to the sign of the acceleration.

43. Formulæ for Uniform Motion. — Let v represent the velocity of a body having uniform motion during t units of time, then vt will be the distance passed over; representing this distance by s, we have $s = vt$. From this equation we derive $v = \dfrac{s}{t}$ and $t = \dfrac{s}{v}$.

44. Formulæ for Uniformly Accelerated Motion. — Let a represent the acceleration and v the velocity at the end of t seconds. Since the velocity changes a units each second, in t seconds the change will be at. If the body starts from rest, the final velocity will be at, or $v = at$. Since the initial velocity is 0 and the final velocity at, the average velocity during the time t will be $\frac{1}{2}(0 + at)$ or $\frac{1}{2}at$. Hence the distance passed over is $s = \frac{1}{2}at \times t = \frac{1}{2}at^2$.

MECHANICS OF SOLIDS.

EXERCISES.

1. If a body has a speed of 60 miles an hour, what is the speed per second expressed in feet?

2. A railroad train, 120 yds. long, passes over a bridge 80 ft. long, at the rate of 30 miles an hour. How long does the train take to pass completely over the bridge?

3. Roemer found that a ray of light took 16 m. 36 s. to cross the diameter of the earth's orbit. The mean distance of the sun from the earth is 92.39 million miles. What must be the velocity of light per second?

4. A body starting from rest moves with an acceleration of 20 ft. per second. What space does it pass over in 6 seconds, and what is the velocity at the end of that time?

5. A train, starting from rest, moves with a uniform acceleration and takes 5 minutes to pass over the first mile. What is the acceleration per second, and what is the final velocity?

6. A body, uniformly accelerated, starts from rest, and passes over a certain space, b, in the first second. Show that it will pass over $15b$ during the following three seconds.

45. Graphic Representation of Velocities. — Uniform motion in a straight line is completely described when its *magnitude* and *direction* are given. Since a straight line possesses these two characteristics, it is evident that a velocity may be represented both in magnitude and direction by a straight line. Thus, if a body has a velocity of 20 metres per second, in a direction parallel to the top edge of this page, then by representing by a line 1 cm. long a velocity of 5 m. per second, the line AB 4 cm. long (Fig. 13), drawn parallel to the top edge of the page, will represent a velocity of 20 m. per second.

Fig. 13.

46. Composition of Velocities. — Many cases of velocities are made up of two or more velocities compounded; for example, when a person is walking about on the deck of a moving ship, his actual, or *Resultant*, velocity in space depends on both the velocity of the ship and his own velocity relative to the ship. The velocities of which the actual velocity is compounded are known as the *Component Velocities*.

47. Cases of Compounding Velocities. — In compounding velocities several distinct cases arise; among them are the following:

First. — *When two velocities in the same straight line are compounded, the resultant velocity is the algebraic sum of the components, and its direction is that of the greater.*

For example: A boat which can be moved in still water at the rate of 5 miles an hour, will have what velocity against a current of 3 miles an hour?

According to the above principle the velocity will be $5 - 3 = 2$ miles in a direction opposite to that of the current.

Second. — *If two adjacent sides of a parallelogram represent two component velocities imparted to a body, their resultant is represented in magnitude and direction by the diagonal of the parallelogram drawn through the intersection of the two adjacent sides representing the velocities.*

This principle is known as that of the *Parallelogram of Velocities*. The following example will make clear its meaning:

If a boat can be rowed in still water at the rate of 5 miles an hour, what will be the resultant velocity if rowed at right angles to a current of 3 miles an hour?

Let a line whose length is 1 cm. be taken to represent the unit of velocity; then a line AB (Fig. 14), 5 cm. long, will represent the velocity due to rowing, and a line, AC, 3 cm. long, drawn at right angles to AB, will represent the velocity due to the current. Completing the parallelogram, its diagonal AD will represent the actual velocity. By the aid of dividers and scale we find AD to be 5.83 cm. long. Hence, the resultant velocity is 5.83 miles per hour, and the direction is that of AD.

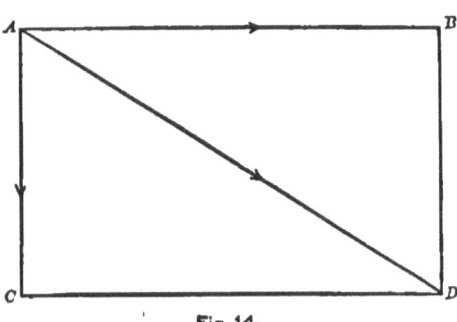

Fig. 14.

When the angle between the component velocities is a right angle, the value of the resultant velocity will be the square root of the sum of the squares of the two component velocities. In the above problem the resultant velocity will equal $\sqrt{5^2 + 3^2} = \sqrt{34} = 5.83$. When the angle is not a right angle, the value can be found approximately by the graphic process illustrated in Fig. 14. When greater accuracy is required, the principles of Plane Trigonometry must be used.

Third. — *When a body has three or more velocities imparted to it simultaneously, the resultant velocity may be found by repeated applications of the foregoing principles.*

For example: If a body has a velocity eastward of 100 feet per second, northward of 80 feet per second, and north-westward of 120 feet per second, what is the resultant velocity?

Let *AB*, *AC*, and *AD* (Fig. 15) represent these velocities on a scale of 80 feet to the inch. Then *AE* is the resultant velocity of *AB* and *AC* (why?), and *AF* of *AE* and *AD* (why?). The length of *AF* in inches, as found by the aid of a scale and dividers, multiplied by 80, will be the value of the resultant velocity.

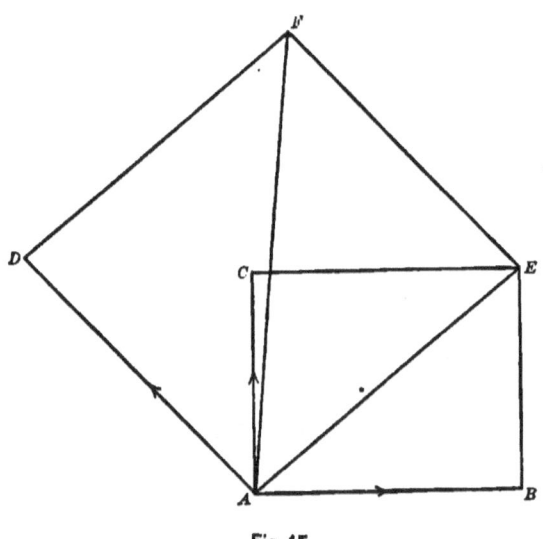

Fig. 15.

48. Resolution of Velocities. — It is often desirable to find two or more velocities which may be substituted for a given velocity without affecting the result. The process is known as the *Resolution of Velocities*, and is evidently the converse of the *Composition of Velocities*. The most important case is the converse of the Parallelogram of Velocities, its solution consisting in finding the sides of the parallelogram whose diagonal represents the given velocity. An inspection of Fig. 16 will make it evident that an infinite number of parallelograms are possible of which

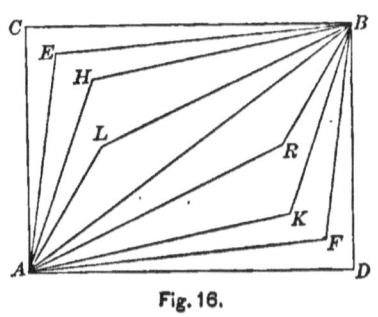

Fig. 16.

AB is a diagonal. Hence the velocity represented by AB can be replaced by AC and AD, AE and AF, AH and AK, etc.

In the application of this principle the direction of the components is usually known. For example, let it be required to resolve a velocity of 40 miles per hour eastward into two velocities at an angle of 60°, one of which shall be south-eastward. Let AB (Fig.17) represent a velocity of 40 miles per hour eastward. Draw AE making an angle of 45° with AB, and AF making an angle of 60° with AE.

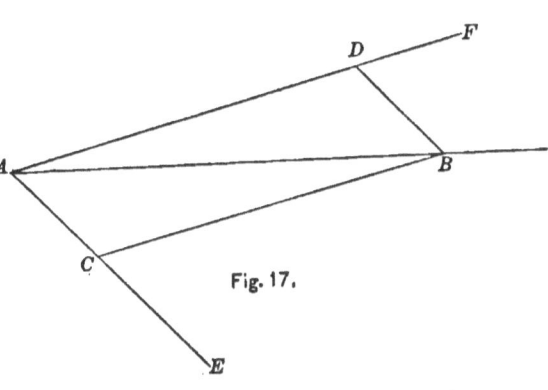

Fig. 17.

Through B draw BD and BC parallel respectively to AE and AF, forming the parallelogram $ACBD$. Hence, the components required are AC and AD, and their numerical equivalents can be obtained by a scale of equal parts.

EXERCISES.

1. When a train is moving with a velocity of 20 miles an hour past a station platform, the conductor throws out a parcel with a horizontal velocity of 20 feet per second in a direction at right angles to the motion of the train. What will be the velocity of the parcel at the beginning of its flight?

2. If a steamer has a velocity of 14 miles an hour due east, and the wind blows with a velocity of 8 miles an hour from the south, what will be the apparent velocity of the wind to one on board the steamer?

3. A particle receives simultaneously two velocities, 100 feet per second north, and 75 feet per second north-east. Find the magnitude and direction of the resultant velocity.

4. A body is acted on simultaneously by three forces, capable of giving it the velocities 50, 70, and 100 feet per second respectively; the angle between the directions of the first two is 60°, and that between the second and third, 90°. Find the magnitude and direction of the resultant velocity.

5. A body has four component velocities, viz.: 60 feet per second east, 70 feet per second north, 80 feet per second west, 90 feet per second south. Find the magnitude of the resultant velocity.

6. A body has two component velocities, 8 and 10, inclined at an angle of 60°. Find the magnitude and direction of the resultant velocity.

7. Two equal velocities are simultaneously impressed upon a body, one toward the north, the other toward the east. The resultant velocity is equal to 10. Find the two velocities, and the direction in which the body will move.

8. A train is moving south-eastward at a velocity of 30 miles an hour. How fast is it moving eastward, and how fast southward?

9. A road is inclined to the horizon at an angle of 30°. What velocity does a carriage have vertically and horizontally, when it moves up the slope at the rate of 8 miles an hour?

10. The sail of a ship is set in such a manner that the wind strikes it at an angle of 30°. If the wind has a velocity of 15 miles an hour, what would be the value of that component perpendicular to the sail?

II. NEWTON'S LAWS OF MOTION.

49. Momentum means *quantity of motion*. It is proportional both to the quantity of matter moving and to its velocity, and is measured by the product of the mass of the body and its velocity. For example: if a body whose mass is 40 pounds is moving with a velocity of 20 feet per second, then its momentum is $40 \times 20 = 800$.

50. The Laws of Motion.—The effect of force in producing alteration of the motions of bodies is fully defined and described in the following propositions, known as *Newton's Laws of Motion:*

I. *Every body continues in its state of rest or of uniform motion in a straight line, except in so far as it may be compelled by impressed force to change that state.*

II. *Change of motion is proportional to the force applied, and takes place in the direction of the straight line in which the force acts.*

III. *To every action there is always an equal action in an opposite direction.*

These laws are of the nature of axioms, incapable of experimental proof. They form the foundation of the science of Kinetics.

51. Discussion of the First Law.—This law asserts that if a body be at rest, and be left wholly to itself, it will forever remain at rest; that if a body be at one moment at rest, and be subsequently found in motion, then it has been acted on by some force; that if a body be in motion and be subsequently found at rest or moving at a different rate or in a different direction, then it has been acted on by a force. Hence, this law teaches that a *Force is that which changes, or tends to change, a body's state of rest or motion.*

52. Discussion of the Second Law. — The meaning of this law is that a force is measured by the change of momentum per second which it produces; that is, $F = Mv \div t$ (49), where F, M, v, and t represent force, mass, velocity, and number of seconds, respectively. By (44) $v \div t = a$. Hence, $F = Ma$. It follows that two forces may be compared by comparing the momenta generated in equal times or by the accelerations imparted to equal masses.

The law shows further that if several forces act simultaneously on a body, each produces its own change of motion, in its own direction, and that forces may be compounded in the same way as velocities. Change of motion of a moving mass means change of momentum.

53. **Units of Force.**—There are two units of force, the *Gravitational*, and the *Dynamic* or *Absolute*. The *Gravitational Unit* is the weight of some standard mass, as the *pound* or *kilogramme*, and is accordingly called the *Pound* or *Kilogramme*. Since the force of gravity varies with the place, these gravitational units of force will vary; and though they are convenient for the measurements of the engineer, they are not suitable for precise investigations. The *Dynamic* or *Absolute Unit of Force* is that force which acting for a unit of time on a unit of mass produces a unit change of velocity. If the unit of mass is the pound and the unit of velocity a foot per second, the unit of force is a *Poundal;* if the unit of mass is the gramme and the unit of velocity a centimetre per second, then the unit of force is a *Dyne*. These units are invariable in value, being independent of the attraction of the earth.

If the force of gravity at New York act on a mass for one second, it has been found that it will impart to it a velocity of 32.16 feet per second; hence, if the mass is one pound, the force of gravity is equal to 32.16 poundals, and the poundal is $\frac{1}{32.16}$ of the earth's attraction for the pound-mass, or nearly half an ounce at that place. In like manner, it appears that the earth's attraction for one gramme at New York is 980 dynes, since 32.16 feet equals 980 centimetres; and one dyne is $\frac{1}{980}$ of the earth's attraction for one gramme-mass.

54. How Forces are Measured.—It has already been pointed out (52) that forces may be compared by comparing the momenta generated in equal intervals of time. The simplest way, however, of measuring a force is by the use of a *Dynamometer*, which consists of a spring, to some part of which is attached a pointer or index, moving in front of a graduated scale. The instrument can be arranged to give its readings in pounds, grammes, poundals, or dynes. The common draw-scale (Fig. 18) is a dynamometer, graduated to pounds and fractions of a pound.

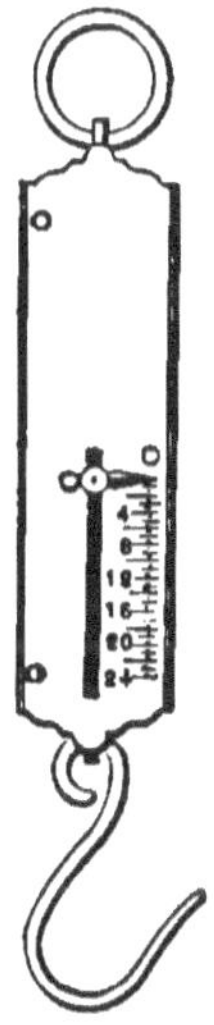

55. Graphic Representation of a Force.— A force is completely specified when there are given: (1) *Its magnitude*, (2) *Its direction*, (3) *Its point of application*, that is, the point of the body at which it acts. Since straight lines possess all these characteristics, it is evident that forces, like velocities (45), may be represented by them. The point from which the line is drawn may represent the point of application; the length of the line, the magnitude of the force; and the direction of the line, the direction in which the force acts. For example, if a line one inch long be used to represent a force of one unit, then a force of 10 units would be represented by a line 10 inches long.

Fig. 18.

ELEMENTS OF PHYSICS.

56. Composition of Forces.—If two or more forces act on a body, and if a single one can be found which will produce the same effect as the two or more forces acting together, it is called the *Resultant*. If two or more forces act on a body in such a manner as to put it in motion, and

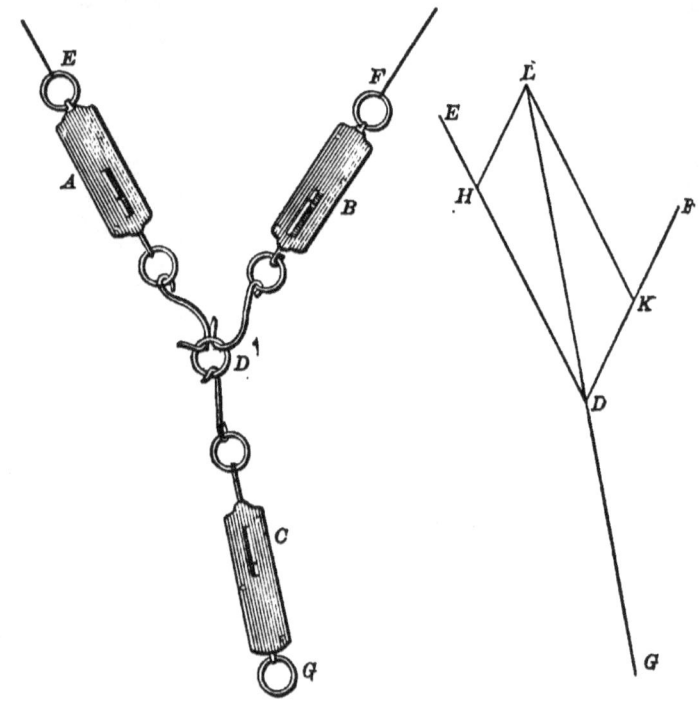

Fig. 19.

a single force can be found which, acting along with these two or more, will keep the body at rest, it is called the *Equilibrant*.

The more important cases of the Composition of Forces are:

First. — *If two concurring*[1] *forces act on a body at an*

---
[1] Concurring forces are those having a common point of application.

angle, the concurrent diagonal of the parallelogram whose adjacent sides represent the directions and the intensities of these forces will represent their resultant.

This is known as the principle of the *Parallelogram of Forces*, and may be verified as follows:

Exp. — Hook three draw-scales, A, B, and C, into a small iron ring, D (Fig. 19). Fasten two of them to stout pins, E and F, in the frame of the black-board. Pull on G, record the readings of A, B, and C, and mark the centres of the rings D and G. Now remove the draw-scales, and draw lines from E, F, and the centre of G, to the centre of D. Let the readings of A, B, and C be 15, 7, and 20, respectively. Using some convenient unit, lay off on DE, DF, and DG, the distances 15, 7, and 20, respectively. Complete the parallelogram $DHLK$, draw the diagonal DL, and find its value by measurement. If the work be carefully done, it will be found that DL and DG are in the same straight line and are equal in value. Hence, the single force DL is the resultant of the forces DH and DK, as it equals DG, their equilibrant.

Second. — If more than two concurring forces act on a body, the resultant may be found by finding the resultant of any two, then of that resultant and a third, and so on till all the forces have been considered. The last resultant will be the one required.

This may be verified experimentally by proceeding as in the first case, using four or more draw-scales.

Third. — The resultant of two parallel forces having the same direction is their sum, and its point of application divides the line joining the points of application of the two components inversely as the magnitudes of those forces.

This principle may be verified by the following experiment:

Exp. — Suspend two draw-scales, A and B, from a suitable frame (Fig. 20). Through their hooks slide a wooden rod. On the side opposite to A and B, hook to the rod a third draw-scale, C. Arrange A and B so that they are parallel; then pull on C, and record the readings of each draw-scale, as well as the distances DK and KH.

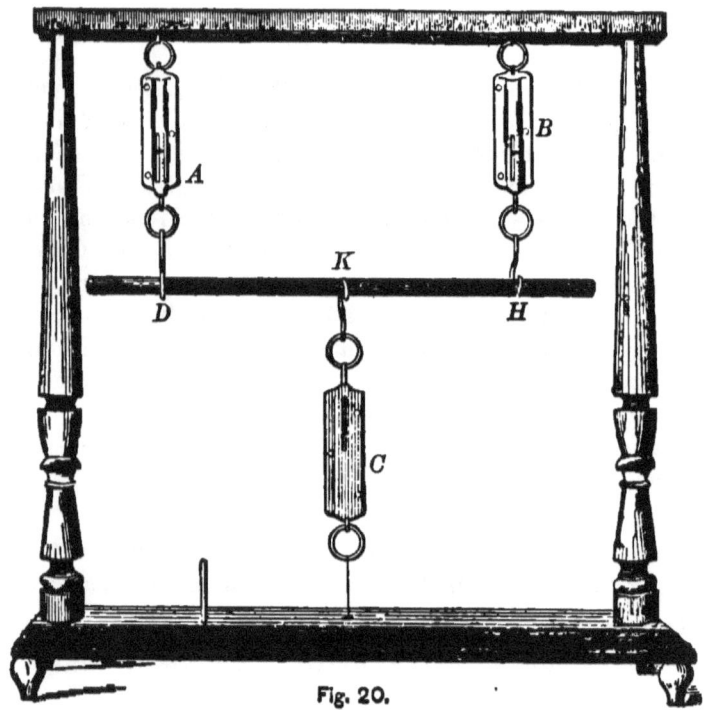

Fig. 20.

If carefully conducted, the experiment will show that $C = A + B$, and $\frac{A}{B} = \frac{KH}{KD}$. A known weight might be substituted for the third draw-scale.

Fourth. — *If two equal parallel forces act at different points on a body and in opposite directions, they produce rotary motion; no single force can replace them.*

Such an arrangement is known as a *Couple*. An illustration is to be seen in the action of the earth on a magnetic needle. If such a needle be displaced from its position of north and south, the earth's magnetism attracts one end of it and repels the other with equal parallel forces, thereby constituting a couple with the effect of merely rotating the needle about an axis, till it returns to its position of north and south.

57. Resolution of Forces. — It is often desirable to replace a single force by two or more forces without changing the effect, or it is required to determine what part of a force is acting in a certain direction. To illustrate: Let it be required to resolve a force of 32 lbs. into two components acting at right angles, one of them to be 12 lbs.

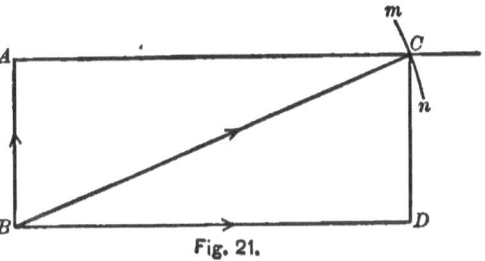

Fig. 21.

The problem is evidently to construct a rectangle with a diagonal of 32 and a side of 12. Draw AB (Fig. 21) 12 units long; at A erect AC perpendicular to AB; with B as a centre and with a radius of 32 units, draw the arc mn cutting AC at C; through C draw CD parallel to AB, and through B draw BD parallel to AC. By (56) it is evident that BA and BD are the components of BC; and hence, BD is the component sought, and its value can be found by means of scale and dividers.

As a second illustration, let a body, W, whose weight is 20 lbs., be supported on the inclined surface, AB (Fig. 22); what force acting parallel to the plane will be necessary to maintain it? Draw WH perpendicular to

AB, WF representing the weight 20 lbs., WE parallel to AB, and FH and FE parallel to WE and WH respectively. Then, WH and WE are components of WF, WH represents the pressure on AB, and WE represents the force urging W toward B. Hence, to maintain W at rest, it will be necessary to apply a force WK equal and opposite to WE.

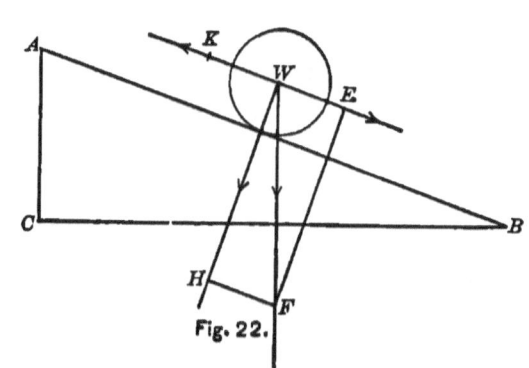

Fig. 22.

EXERCISES.

1. A force of 100 pounds at New York equals how many poundals?
2. A constant force acting on a mass of 15 grammes for 4 seconds gives it a velocity of 20 centimetres per second. Find the force in dynes.
3. Find how many dynes in a poundal.
4. What is the force in poundals that in 10 minutes produces a velocity of a mile a minute in 100 lbs.?
5. A mass of 15 lbs. lying on a smooth, flat table is acted upon by a force of 60 poundals; what velocity will it have at the end of 2 seconds?
6. A force of 30 dynes acts for 12 seconds upon a body resting on a smooth, horizontal plane, and imparts to it a velocity of 120 centimetres per second; what is the mass of the body?
7. An 18-ton truck is moving at the rate of 30 miles an hour; what is its momentum?
8. Compare the momentum of a 15-lb. cannon-ball moving at the rate of 300 feet per second, with that of a 3-oz. bullet which has a velocity of 420 yds. per second.

9. Two balls have equal momenta. The first weighs 100 kilos., and moves with a velocity of 20 metres a second. The other moves with a velocity of 500 metres a second. What is its mass?

10. Two forces, 60 and 91 lbs., act at an angle of 90°. Find their resultant.

11. Two forces act at an angle of 60°. Their resultant is 40 lbs., and one of the forces is 25 lbs. Find the other force.

12. Forces 20, 30, 40, act on a particle; the angle between the directions of the first two is 45°, and that between the directions of the last two is 60°. Find the magnitude of the resultant.

13. A weight of 25 kilogrammes is suspended by two strings, inclined to the vertical at 30° and 60°. Find the tension of each string.

14. Four forces of 24, 10, 16, 20 dynes act on a particle, the angle between the first and second being 30°, between the second and third 90°, and between the third and fourth 120°. Find the magnitude of the resultant.

15. Resolve a force of 25 lbs. into two components at right angles, one of the forces to be 10 lbs.

16. Find the point of application of the resultant of two forces, 7 and 13 units, acting in parallel lines 40 inches apart and in the same direction.

17. A bar 8 metres long is supported in a horizontal position by props placed at its extremities; where must a weight of 100 kilogrammes be hung so that the pressure on one of the props shall be 20 kilogrammes?

18. Find the points of application of two forces acting along parallel lines one metre apart, so that one may be four times the other and their resultant 120 kilogrammes.

58. Discussion of the Third Law.—If one exerts a downward pressure on the table with his hand, the table presses the hand upward with an equal force. A magnet attracts a piece of iron with a certain force and the iron attracts the magnet with an equal force acting in an opposite direction. A horse drawing a body by means of a rope is pulled backward by a force equal to that with which the body is drawn forward. These illustrations suggest that in every action of force, two forces are apparently involved; or rather that the term *Force*, as usually employed, designates only one part of the mutual action between bodies. To designate the entire action, the term *Stress* is employed, and it receives different names according to the aspect under which it is studied, as *pressure, tension, attraction*, etc.

59. Reflected Motion is the motion produced in a body by the reaction between it and a second body against which it strikes. For example, let a highly elastic ball be thrown in the direction of *AB* (Fig. 23), against a surface

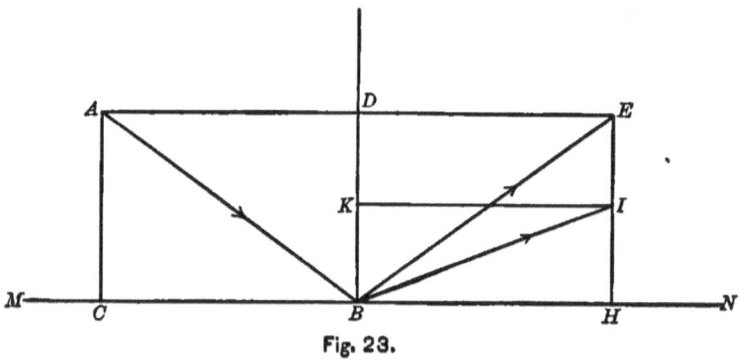

Fig. 23.

MN, and let *AB* represent the magnitude of the incident force. This may be resolved into two components, *DB* and *CB*, the former acting normally to the surface *MN*, and the latter acting in line with the surface. If the ball and the plane are each perfectly elastic, then the reaction at *B* will equal the normal component *DB*; and as *CB* is not affected in any way, it follows that after the collision there are two forces acting at *B*, *BD* and *BH*, equal respectively to *DB* and *CB*. The resultant of these two forces is *BE*, a force evidently equal to *AB* and making with *BD* an angle equal to that made by *AB*.

If the ball and the plane are not perfectly elastic, the force due to reaction will be something less than *BD*, as

BK, and the resultant *BI* will be less than *BE*, but will make a larger angle with *BD*. When the elasticity is *nil*, *BD* will be zero; and as *BH* will be the only force acting, the ball will move in obedience to that force and roll along the plane.

The Angle of Incidence is the angle between the direction of the incident body and the normal to the surface at the point of incidence. *The Angle of Reflection* is the angle between the direction of the reflected body and the normal. In Fig. 23 the angle *ABD* is the *angle of incidence*, and *EBD* that of *reflection*.

60. Law of Reflected Motion. — An examination of the preceding discussion reveals the following law:

For perfectly elastic bodies, where the force of restitution equals the force of compression, the angles of incidence and reflection are equal to each other; but for imperfect elasticity, the angle of reflection is larger than the angle of incidence.

Exp. — Cut from a board a semicircular piece having a radius of about 24 cm. (Fig. 24.) At the centre of the semicircle fasten, with its face in the diameter, a small rectangular piece of polished stone or iron. Draw radial lines, dividing the circle into ten-degree spaces. Suspend from a suitable support a glass ball about 2 cm. in diameter, so as just to touch the polished surface, exactly over the centre of the semicircular base. Stand a common chalk crayon on one of the radial lines; draw off the ball, and release it successively over different radial lines in such a manner that it will strike the polished surface exactly over the centre of the base. Notice that the reflected ball overturns the crayon only when it falls along that radial line which makes with the normal to the reflecting surface an angle equal to that between the

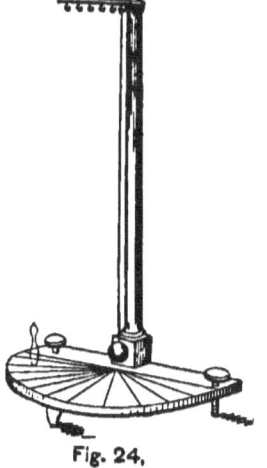

Fig. 24.

normal and the radial line on which the crayon stood, proving the equality of the angles of reflection and incidence.

If a lead ball be used, it will be found that the crayon is overturned only when the angle of incidence is considerably smaller than that of reflection, a conclusion in harmony with the law for imperfectly elastic bodies.

III. WORK AND ENERGY.

61. Work. — When a body is moved by the action of force upon it, the force is said to do *Work* on the body. For example, steam exerts pressure on the piston in the cylinder of an engine, causing it to move. The expansive force of the steam does work on the piston in overcoming its resistance to motion. If there had been no motion of the piston, no work would have been done. In merely supporting a body, no work is done, since no motion is produced. The force of gravity does no work on a stone resting on the ground, yet it causes a pressure between the stone and the earth.

62. Units of Work. — To estimate the work done by a force on a body, some unit of work must be selected. Four such units are in use: the *Foot-pound*, the *Foot-poundal*, the *Kilogramme-metre*, and the *Erg*. A *Foot-pound* is the work done by a force of one pound working through a space of one foot. This is the one in general use among English-speaking engineers. As a unit it is open to the objection that it is variable, owing to the variation of the pound with the latitude (53). The *Foot-poundal* is an absolute unit, being the work done by a force of one poundal working through a space of one foot. Since there are 32.16 poundals in the weight of one pound mass at New York, then there are 32.16 foot-poundals in the foot-pound at that place.

MECHANICS OF SOLIDS. 45

In the metric system, the corresponding units are the *Kilogramme-metre* and the *Erg*. The former is the amount of work done by a force of one kilogramme working through a space of one metre; the latter is the amount of work done by a force of one dyne working through a space of one centimetre. Since there are 980,000 dynes in the weight of one kilogramme-mass at New York, and 100 centimetres in a metre, then in a kilogramme-metre there are 98,000,000 ergs at that place.

63. **Power.** — The rate at which work is done is called *Power* or *Activity*. The unit of power in general use in the English gravitational system is the *Horse-Power;* it is the rate of doing work equal to 550 foot-pounds per second, or 33,000 foot-pounds per minute. In the absolute system the unit of power is the *Watt*, which equals work done at the rate of 10^7 ergs per second. One horse-power is equivalent to 746 watts. Therefore to convert horse-power into watts, multiply by 746.

64. **Work Computed.** — Since a unit of work is done when unit force displaces a body through unit distance, then 5 units are done by a force of 5 units displacing a body through a unit distance, and 5 × 2 units of work are done if this force produces a displacement of 2 units of distance. Hence, *the amount of work done by a force is expressed by the product of the number of units of force by the number of units worked through.* Let F equal the number of units of force, and s the space moved over; then the *work done* equals $F \times s$.

Since the resistance overcome has the same numerical measure as the force acting (58), the work can be computed by multiplying the resistance by the space, or *work* done equals $F \times s$, in which F is the measure of the resistance overcome. When a body is lifted in opposition to gravity, F is evidently the weight; in moving a body in any other direction F must be determined either by computation or by experiment.

65. Energy.—Experience teaches that a body in motion can impart motion to a second body. For example, the moving bat in the hands of the ball-player may impart motion to the ball, that is, it does work on the ball. Hence, *a body in motion can do work on other bodies.*

Again, if we lift a weight from the floor to the table, we do work upon the weight. Now if we attach a cord to this weight, pass it over a pulley or grooved wheel, fasten the free end to a second weight somewhat smaller and remove the supporting table, the weight falls to the floor, and in so doing raises from the floor the smaller weight, that is, does work upon it. If the table had been higher, or the first weight had been heavier, then more work would have been done in placing it on the table, and it could have done more work in falling to the floor. Hence, *a body by virtue of its position can do work.*

Energy is the capacity of doing work. When a system acquires the capacity of doing work because of the work done on it, it is said to possess energy.

66. Types of Energy.—As seen by the illustrations given in the last section, there are two types of energy, *one due to the motion of the body and the other due to its relative position.* The former is called *Kinetic Energy* and the latter *Potential Energy* or *Energy of Position or Stress.*

67. Kinetic Energy may be defined as the capacity of doing work possessed by a body by virtue of its motion.

If a ball be thrown vertically upward, it possesses *kinetic energy* at the beginning of its journey, because it has motion. If intercepted by a second body, it would impart motion to that body, that is, it would do work upon it. If the ball be left to itself, it gradually loses its motion and finally comes to rest. At this moment it possesses no kinetic energy, but, as will be explained later (70), the energy has been transformed into the other type.

68. **Potential Energy** is the energy possessed by a body by virtue of its position with reference to some other body, or by virtue of the relative position of its parts. Thus, when a stone has been lifted to a certain height above the earth's surface, work has been done in overcoming the attraction existing between the earth and the stone. If the stone is allowed to fall to the earth, it is able to do a certain amount of work. Hence, the stone in its elevated position possesses *Potential Energy*, or, as it is sometimes called, *Energy of Position*. In coiling up a spring, or bending a bow, work is done in distorting the body, or placing it under stress; that is, the relative position of the parts of the body is changed; and it possesses, by virtue of this distortion, potential energy.

69. **Energy Measured.** — It is evident from the definition of energy that it is not only a measurable quantity, but also that it is a quantity to be measured by the same units as those used in measuring work.

In uniformly accelerated motion (44) $s = \frac{1}{2}at^2$ and $v = at$. Eliminating t, we have $v^2 = 2as$. Hence, $\frac{1}{2}Mv^2 = \frac{1}{2}M \times 2as = Mas$. But $Ma = F$ (52). Therefore $F \times s = \frac{1}{2}Mv^2$.

Since $F \times s$ is a measure of the work done (64) and

hence of the energy expended, or to be expended, then the Kinetic Energy in absolute units, foot poundals, or ergs is expressed by $\tfrac{1}{2}Mv^2$.

In the latitude of New York, foot-poundals can be reduced to foot-pounds by dividing by 32.16, and ergs to kilogramme-metres by dividing by 98,000,000 (62).

70. Transformations of Energy. — It has already been said that if a ball be thrown vertically upward, it gradually loses its motion, and hence its kinetic energy, but it acquires energy of position. If the ball be allowed to fall, it regains its original motion, and hence its kinetic energy; but it loses its potential energy. When the ball strikes the ground its motion ceases, and its energy is apparently destroyed. Extended experiments have demonstrated that both the ground and the ball are warmer after the impact than before, by an amount bearing a fixed relation to the amount of energy used up. It will be seen in Chapter IV. that heat is due to the motion of the molecules of a body, and that the kinetic energy of the body as a whole has been divided up between the molecules of the body and the ground, thereby increasing their kinetic energy or heat. The energy of a heat-engine is derived from the heat evolved by the combustion of the fuel, and this fuel may be regarded as representing a certain amount of the energy of the sun (its heat), transformed into potential energy in the cell of the growing plant.

71. Conservation of Energy. — Energy resembles matter in being unchangeable in amount. It cannot be given to one body without taking it from another. The energy of the flying ball came from the arm which threw it, and that of the rapidly moving locomotive from the steam in the engine cylinder. In all transferences and transformations of energy there is no loss, that is, the total energy of the universe is constant.

72. Matter and Energy.—"All that we know about matter relates to the series of phenomena in which energy is transferred from one portion of matter to another till in some part of the series our bodies are affected, and we become conscious of a sensation. We are acquainted with matter only as that which may have energy communicated to it from other matter. Energy, on the other hand, we know only as that which in all natural phenomena is continually passing from one portion of matter to another. It cannot exist except in connection with matter."[1]

EXERCISES.

1. What causes the "kick" of a gun? Why is it less in the case of a heavy gun?

2. Two bodies, weighing 50 pounds and 75 pounds respectively, have the same momenta. If the first has a velocity of 1,000 ft. per second, what velocity has the second?

3. Of what horse-power is an engine which can raise 20 tons in 5 minutes to the height of 50 feet?

4. How many gallons of water can a 100 H.-P. engine raise 200 feet high in 10 hours? Assume that a gallon of water weighs 8 lbs.

5. A shot of 30 pounds is fired from a gun of 3 tons, and leaves it with a velocity of 1,500 feet per second; find the velocity of the gun's recoil.

6. A mass of 1,000 pounds, moving with a velocity of 500 feet per second, possesses how much energy?

7. What is the kinetic energy of a mass of 500 kilos. after it has fallen through a vertical distance of 8 m.? (93.)

8. James Watt found that an English dray horse could travel at the rate of 2.5 miles an hour, and at the same time raise a weight of

---

[1] Maxwell's "Matter and Motion," page 168.

150 lbs. by means of a rope led over a pulley. How much work was done per minute?

9. If a cannon-ball were discharged from the rear end of an express-train, directly along the track, at the same rate as the train is moving forward, what would be the motion of the ball relative to the ground?

10. Find the work done by a force of 100 dynes acting through a distance of 25 metres. Express the result in ergs.

11. Express in foot-poundals the work done in raising a weight of one ton through a vertical height of 5 yards.

12. What is the potential energy in foot-pounds of a gallon of water in a reservoir at the height of 200 feet above the ground?

13. What is the energy of a mass of 5 kilogrammes moving with a velocity of 50 metres per second?

14. A steam-engine supplies 1,000 houses with 100 gallons of water each, working 12 hours per day; if the mean height to which the water has been raised is 80 feet, at what rate does the engine work?

IV. GRAVITATION.—CENTRE OF MASS.—STABILITY.

73. Universal Gravitation. — The earth attracts to it all objects on its surface giving them *Weight*. Sir Isaac Newton was the first to establish the proposition that this attractive force is not limited by distance, but that every particle in the universe attracts every other particle. This mutual action is known as *Universal Gravitation*.

74. Law of Attraction. — *The attraction between two bodies varies directly as the product of their masses, and inversely as the square of the distance between their centres of mass* (77).

To illustrate: If the attraction between two units of mass separated by a unit of distance be a, then a body which contains m units of mass will attract a unit mass at unit distance with a force of ma, and its attraction for a body of n units of mass will be mna. In like manner, it may be shown that the attraction of the second body for the first is mna. If the distance between the bodies be increased to 2 units, then the attraction will be $\frac{1}{4}$ of what it was before; if to 3 units, $\frac{1}{9}$; if to d units, $\frac{1}{d^2}$ of mna, that is, the attraction is $\frac{mna}{d^2}$.

75. Gravity is the attraction of the earth for other bodies. Its measure in every case is the *Weight* of the body. The tendency of any body to fall toward the earth is due to the mutual attraction between that body and the earth. The path described by a falling body is a *Vertical Line*, and a line or plane perpendicular to it is said to be *Horizontal*. Vertical lines point toward the earth's centre; but those drawn through neighboring points may be considered as parallel without sensible error, owing to the great distance of their point of meeting. The direction of the vertical line at any point can be determined by suspending a weight by a string passing through the point. Such a device is called a *Plumb-line*.

76. Law of Weight. — Since the earth is not a perfect sphere, it follows from the law of attraction that the weight of an object at one place may differ from its weight at another place. *For bodies above the surface, the weight varies inversely as the square of the distance from the centre; for bodies below the surface, the weight would vary as the distance from the centre if the earth's density were uniform.*

77. Centre of Mass. — If the particles of a body are acted on by a number of parallel forces proportional to their masses, the point of application of the resultant of these forces is the *Centre of Mass* of the body. Since the attraction between the earth and any body on its surface may be considered as such a system of parallel forces, a force to each particle of the body, the point of application of the resultant (the weight) is the *Centre of Mass* or *Gravity*. This point is also known as the *Centre of Inertia;* because when a body is acted on by any force, there is, owing to inertia, an equal reaction in the opposite direction on the part of each particle, which is equivalent to a series of parallel forces whose resultant would have its point of application at the *Centre of Mass*.

It is evident from what precedes that the mass as well as the weight of a body may be considered as concentrated at the centre of mass, and that any force acting on a body at its centre of mass will produce a motion of translation; but if it acts at any other point, it forms a couple with the reaction at the centre of mass, and this produces a rotary motion.

78. Equilibrium. — A body is in *Equilibrium* when the resultant of all the forces acting upon it is zero. In the case of a pivoted body, acted on by gravity alone, it can be in equilibrium only when *the vertical line through the centre of mass passes through the point of support*. For let A be a body whose centre of mass is C (Fig. 25), and let B be the pivot or the point of support. Represent the weight by CE. When CE does not pass through B, it is evident that it can be resolved into two components, CD and CF, the former producing pressure on B in the direction of BD, the latter causing C to move

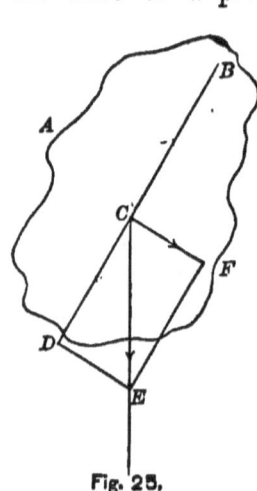

Fig. 25.

to the right. As C approaches the vertical through B, CF will decrease, becoming zero when CE coincides with that vertical. When CE passes through B there can be no motion, as the case becomes one of two equal, concurrent, and oppositely directed forces whose resultant is zero.

In the case of a body supported on a surface, to be in equilibrium it must have *the vertical line through its centre of mass fall within its base.* When this vertical line falls without the base, the body will overturn, for the supporting force does not act opposite to the weight, but parallel to it, forming a couple. If the body is supported on legs, as, for example, a chair, the base is the polygon formed by the lines connecting these points of support.

79. Kinds of Equilibrium. — **Exp.** — Fill a round-bottomed Florence flask one-quarter full of shot, filling the remaining space with paper to keep the shot in place (Fig. 26). Now, tip the flask over, and notice that after a few oscillations it returns to its original upright position. If the experiment be repeated with a similar but empty flask, it will be found impossible to make it maintain,

Fig. 26.

unsupported, any other position than a recumbent one. In this position, however, we may roll it about at pleasure, and it will maintain any position in which both top and bottom rest on the supporting plane.

A study of this experiment reveals the following facts: (1) The centre of mass is nearer the supporting plane in the loaded flask than in the empty one, provided both are erect. (2) In overturning the loaded flask the centre of mass is raised and at the same time the vertical line through it is thrown outside the point of support in such a manner that the effect of gravity is to make it return to its original position; in overturning the empty flask, the centre of mass is lowered and the vertical line through it is brought outside the base in such a way that the effect of gravity is to prevent the flask's assuming a vertical position. (3) In rolling about, the centre of mass of the overturned flask is neither raised nor lowered. Hence, we have three kinds of equilibrium. *Stable*, when the centre of mass is raised by overturning the body; *Unstable*, when it is lowered; and *Neutral*, when it is neither raised nor lowered.

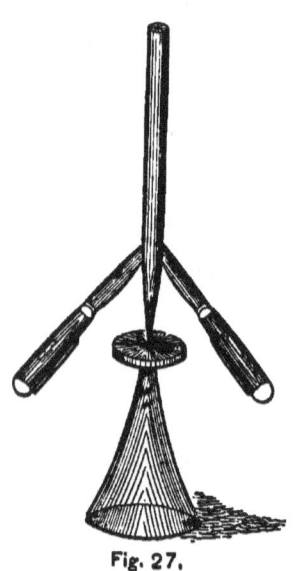

Fig. 27.

80. **Illustrations.** — The rocking-horse, the swing, and the rocking-chair are constructed in such a manner that when moved the centre of mass is raised, and hence they return to their original position, being cases of stable equilibrium. A lead-pencil will not stand on its point because the slightest displacement lowers the centre of mass and the equilibrium is unstable. By inserting two knives with heavy handles as shown in Fig. 27, the centre of mass of the combination is brought below the point of support, and any tipping over of the pencil raises the centre of mass, the knives having changed the case from one of unstable to one of stable equilibrium.

It not infrequently happens that two kinds of equilibrium are illustrated in the same object, as in the case of

MECHANICS OF SOLIDS. 55

a coin placed on edge, which is in neutral equilibrium for displacements in its own plane and in unstable equilibrium for displacements in any other direction.

81. The Stability of a body is measured by the work required to upset it, or by the product of its weight by

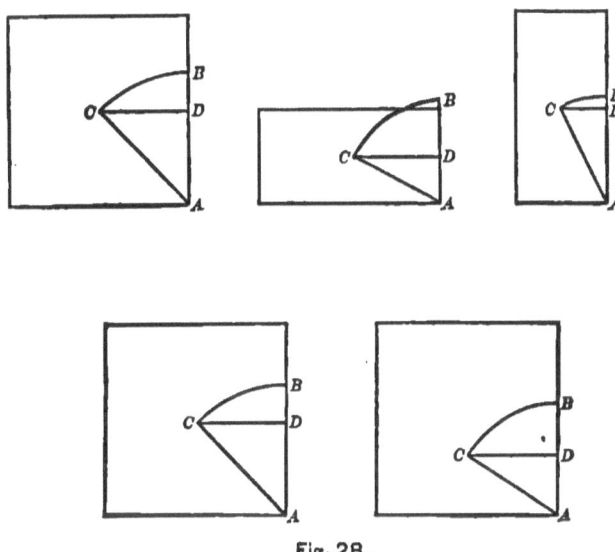

Fig. 28.

the difference between the distances AC and AD in the figure.

An inspection of Fig. 28 will make it evident that *the stability is diminished by raising the centre of mass and is increased by lowering it or by enlarging the base.* C represents the centre of mass, A the point about which it is overturned, and BD the height through which the centre of mass is moved, or the weight lifted, in order to bring it over A, where the condition is one of unstable equilibrium. BD therefore measures the stability, if the masses are considered equal.

82. Applications. — In loading a wagon with an assortment of articles, the heavy ones are placed at the bottom of the load, in order that the centre of the mass may be kept as low as possible. Quadrupeds walk much younger than children, owing to the supporting base being much larger, thereby increasing the stability. The art of the rope-dancer consists in keeping his centre of mass over the supporting rope. The bases of stands are frequently made of some heavy material, in order to keep the centre of mass as low as possible. The boy on stilts finds it difficult to maintain his position because of the height of the centre of mass above the base, and the decrease in the size of the base.

EXERCISES.

1. How far below the surface of the earth will a 10-lb. ball weigh only 4 lbs.?
2. A body at the earth's surface weighs 900 pounds; what would it weigh 8,000 miles above the surface?
3. What effect would it have on the weight of a body to double both its mass and also that of the earth?
4. How far below the surface of the earth must an avoirdupois pound-weight be placed, in order to weigh an ounce?
5. What would a body weighing 550 lbs. on the surface of the earth weigh 3,000 miles below the surface?
6. If a body falls 16 ft. in a second at the surface of the earth (radius = 4,000 miles), how far would it fall in one second at the distance of 240,000 miles from the earth's centre?
7. The mass of the sun is 330,000 times that of the earth, and the radius of the sun is 440,000 miles. Find how far a body will fall in one second on the sun, if the earth's radius is 4,000 miles?
8. Why does a person lean forward in climbing a hill?
9. Why is a pyramidal-shaped structure very stable?
10. Why can an old man walk better with the aid of a cane?

MECHANICS OF SOLIDS. 57

11. Why does a ship without a regular cargo carry a ballast of coal, stone, salt, or some heavy material?

12. If a body weighs 100 lbs. on the earth, how much will it weigh on the sun, the sun's mass being 330,000 times that of the earth, and its radius 110 times that of the earth?

V. CURVILINEAR MOTION.

83. **How Produced.** — If a ball attached to the end of a string be whirled around the hand, it will be felt to pull on the string as if endeavoring to break away, that is, there is a constant tension on the string as the ball moves in a circular path. There are evidently, then, two forces involved in the phenomenon, first, that which sets the ball in motion, known as the *Tangential*, and secondly, the *Centripetal*, seen in the tension of the string which deflects the ball from what would otherwise, in obedience to the first law of motion, be a rectilinear path.

84. **The Centrifugal Force** is the resistance offered by the body, due to its inertia, to the deflection from a rectilinear path. It is merely the reaction of the moving body against the Centripetal Force. These two forces are, therefore, but different aspects of the stress along the radius of the curved path described by the body.

85. **Magnitude of these Forces.** — If the mass of a body moving in a circular path be represented by M, the velocity by v, the radius by r, and the acceleration due to gravity by g, it is easily shown that the *Centripetal* or *Centrifugal Force* $= \dfrac{Mv^2}{r}$ in absolute units and $\dfrac{wv^2}{gr}$ in gravitation units. To illustrate: let a body of 10 pounds move in a circle of 5 feet radius with a velocity of 20 feet per second, then the pull on the cord, confining it to the

centre $= \dfrac{10 \times 20^2}{5} = 800$ poundals $= \dfrac{800}{32.2} = 24.85$ pounds at New York.[1]

86. Laws Governing Centrifugal Force. — An inspection of the formula just given suggests the following laws:

I. *The centrifugal force varies as the mass.*

II. *The centrifugal force varies as the square of the velocity.*

III. *The centrifugal force varies inversely as the radius.*

87. The Laws Illustrated. — That the centrifugal force increases with the velocity of rotation is illustrated by the bursting of grindstones and balance wheels when run at too high a speed. A boy's sling illustrates not only the increase of centrifugal force with an increase of velocity, but also the fact that centrifugal force is affected by the length of the radius. On releasing one of the strings, the stone flies off in a line tangent to the curved path described by the sling. The overturning of a carriage, if the driver urges the horses to a high speed when rounding a corner, is also an instance of centrifugal force varying with velocity.

An ingenious application of the tendency of matter,

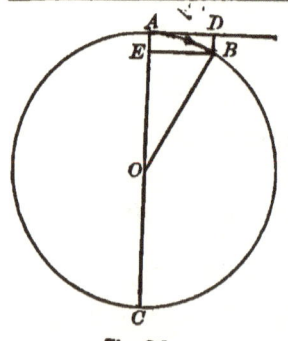

Fig. 29.

[1] Let a ball move uniformly around in a circular path, traversing the distance AB (Fig. 29) in the time t. BD, or AE, will be the deflection of the ball from a straight line due to the centripetal force F, acting toward O. Let a be the acceleration due to this force. Then $AE = \frac{1}{2}at^2$. By Geometry $\overline{AB}^2 = AE \cdot AC$, and by Art. 43, $AB = vt$. By substitution $v^2t^2 = \frac{1}{2}at^2 \cdot 2r$, and hence $v^2 = ar$ and $a = \dfrac{v^2}{r}$. Now $F = ma$ (52). Hence $F = \dfrac{mv^2}{r}$.

when moving in a curved path, to leave that path from inertia is seen in the centrifugal machines employed in laundries for drying clothes, in sugar refineries for removing molasses from the sugar, and in apiaries for extracting honey from the comb. Centrifugal force accounts for the opening out of the balls of a Watt's governor on a steam-engine on an increase of speed, thereby regulating the supply of steam.

The figure of the earth is an oblate spheroid, the diameter through the poles being less than that through the equator. This flattening at the poles was undoubtedly caused by centrifugal force when the earth was cooling down from a liquid to its present solid state.

VI. ACCELERATED MOTION. — FALLING BODIES.

88. Accelerated Motion. — It has been shown in Art. 44 that the relations between *velocity, time, space,* and *acceleration*, in the case of a body having uniformly accelerated motion, are expressed by the formulæ $v = at$ and $s = \frac{1}{2}at^2$. To find the distance passed over during any unit of time, it will be necessary to subtract the distance passed over in $t-1$ units from that passed over in t units. Representing this distance by s', we have $s' = \frac{1}{2}at^2 - \frac{1}{2}a(t-1)^2 = \frac{1}{2}a(2t-1)$.

89. Laws. — From the foregoing formulæ we derive the following:

I. *The velocity at the end of any unit of time is equal to the acceleration multiplied by the number of units of time.*

II. *The acceleration is equal to twice the space passed over during the first unit of time; for if $t = 1$, then $s = \frac{1}{2}a$.*

III. *The distance passed over during any single unit of time is equal to one-half of the acceleration multiplied by one less than twice the number of units of time.*

IV. *The total distance passed over is equal to one-half of the acceleration multiplied by the square of the number of units of time.*

90. Experimental Proof. — The laws of accelerated motion were first verified experimentally by Galileo. His method consisted in reducing the velocity of falling bodies by making only a part of the force of gravity effective in producing motion. In Art. 57 it was shown that if a body rests on an inclined plane only a part of the force of gravity is effective in moving the body down the plane; hence the acceleration is less than that of a body falling freely, and the position of the moving body at the end of any unit of time can be easily marked.

Exp. — Select a straight plank about 16 ft. long, and tack lengthwise of it a straight wooden strip about ½ in. square. To the lateral faces of this strip fasten strips of brass or zinc 1 in. wide, thus making a grooved track, which should be straight and free from inequalities. Raise one end of the plank twelve inches higher than the other. The average of repeated trials will show that an iron ball will roll down this plane a distance of about *one* foot during the *first* second, *four* feet during *two* seconds, *nine* feet during *three* seconds, and *sixteen* feet during *four* seconds. These results may be tabulated as follows:

No. of Seconds.	Total Distance.	Distance during each Second.	Velocity added per Second, or Acceleration.
1	1 ft. $= 1^2 \times 1$	1 ft. $= (2 \times 1 - 1)\, 1$	. . .
2	4 " $= 2^2 \times 1$	3 " $= (2 \times 2 - 1)\, 1$	$2 = 2 \times 1$
3	9 " $= 3^2 \times 1$	5 " $= (2 \times 3 - 1)\, 1$	$2 = 2 \times 1$
4	16 " $= 4^2 \times 1$	7 " $= (2 \times 4 - 1)\, 1$	$2 = 2 \times 1$

An inspection of the above table shows that the total

distance passed over at the end of any second is equal to the square of the number of seconds times the distance passed over the first second ; that the distance during any second is equal to one less than twice the number of seconds times the distance passed over the first second ; and that the distance passed over during the first second is one-half of the acceleration or gain in distance each second.

91. Falling Bodies. — It has already been pointed out that bodies fall to the earth from the attraction of gravity, and that the intensity of this force is affected by distance; that is, the attraction between a falling body and the earth increases as the body approaches the earth. Unless the body falls from a very great height, it will be found that the increase in the attraction is so small that no sensible error is committed by regarding the force of attraction as constant. Hence, if we neglect the resistance of the air, the motion of a falling body will be uniformly accelerated, and the laws for falling bodies will be those for uniformly accelerated motion.

92. The Acceleration due to Gravity is denoted by g, and varies with the latitude, owing to the earth's rotation and its lack of perfect sphericity. The value is least at the equator and greatest at the poles. For the latitude of New York its value is 980 cm. or 32.16 ft.

93. Laws of Falling Bodies. — From what has preceded it is evident that the laws of falling bodies may be derived from those for uniformly accelerated motion by substituting g for a, which gives $v = gt$, $s' = \frac{g}{2}(2t - 1)$, and $s = \frac{1}{2}gt^2$ in which t denotes seconds.

94. Deductions from these Laws. — FIRST, If $t=1$, then $v=g$, that is, the velocity of a body falling freely from a state of rest for one second is equal to the acceleration. SECOND, If $t=1$, then $s=\tfrac{1}{2}g$, that is, the total space passed over in one second by a body falling freely is one-half of the acceleration. THIRD, If the value $t=\dfrac{v}{g}$ be substituted in $s=\tfrac{1}{2}gt^2$, we have $s=\tfrac{1}{2}g\cdot\dfrac{v^2}{g^2}=\dfrac{v^2}{2g}$, from which $v=\sqrt{2gs}$, showing that the velocity acquired by a falling body varies as the square root of the height. FOURTH, It will be noticed that the formulæ for falling bodies contain no expression for the mass of the body. Hence we infer that the mass of a body does not affect its time of falling any given distance. Galileo was the first to point out this truth and to demonstrate it experimentally by letting unequal masses fall from the top of the leaning tower of Pisa.

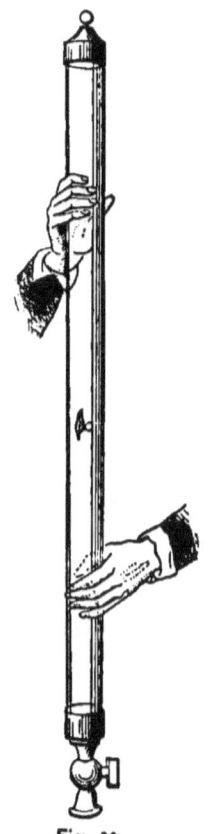

Fig. 30.

Exp. — A glass tube, about four feet long (Fig. 30), is closed at one end, and to the other is fitted a stop-cock; within the tube place a shot and a pith-ball of equal size. Hold the tube in a vertical position, then suddenly invert it; the shot reaches the opposite end first. Now connect the tube to an air-pump, and exhaust the air as perfectly as possible. On inverting the tube, the balls will reach the opposite end nearly together.

The inference from this experiment is that, if the resistance of the air could be wholly eliminated, the time of falling would be independent of the mass. If, however, air be present, the body which has the smaller surface in proportion to its mass will fall the faster, since it suffers less resistance.

The resistance offered by the air to bodies falling through it is illustrated by the case of a mass of water falling over a high precipice, the water being divided into spray before reaching the bottom. In a vacuum water falls like a solid, as is shown by the *Water Hammer* (Fig. 31), an instrument consisting of a closed tube of thick glass, half full of water, the air having been expelled by boiling the water before sealing the tube.

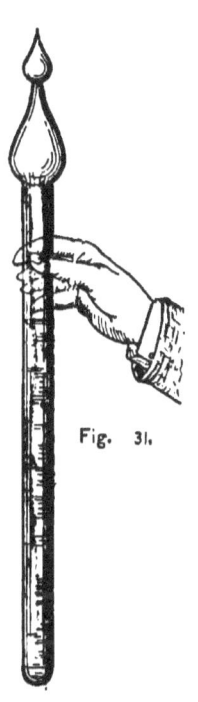

Fig. 31.

95. Bodies Thrown Upward. — If a body be thrown vertically upward, its velocity will be diminished each second by g; hence, neglecting the resistance of the air, the time of the ascent will be found by dividing its initial velocity by g; that is, $t = \dfrac{v}{g}$. It also follows that the times of ascent and of descent are equal, and that a body returns to its starting-point with a velocity equal to that with which it set out.

VII. THE PENDULUM.

96. The Pendulum consists of a body supported so as to move to and fro about a fixed point. In investigating its theory, use is made of an ideal pendulum, which consists of a material particle suspended by a weightless string. Such a pendulum can exist only in the imagination, but its properties can be approximately determined by suspending a small lead ball by a fine thread.

97. Its Motion Explained. — Let N (Fig. 32) be a particle attached by the thread AN to A. If left to itself it takes the position of the vertical through A.

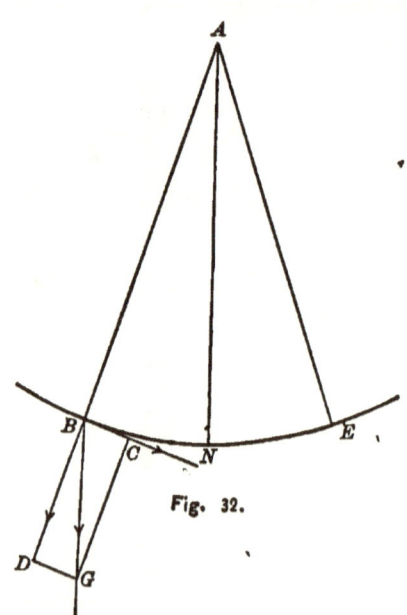

Fig. 32.

If drawn aside to the position B and released, it will move toward N, because of the component BC of the force of gravity BG (57). As it approaches N, this component grows smaller (the student should convince himself of this by drawing a figure with B nearer to N), and is nothing at N. The particle in falling from B to N is under the influence of a continuous force (8), and therefore its motion is accelerated (42), and its kinetic energy continually increased, reaching its greatest value at N. On account of this motion it will pass through N, and rise to E as far on the other side of N as B was, provided there is no resistance from the air. In moving from N to E it is opposed by the effective component of gravity, which passes successively through the same values as in going from B to N, but in reverse order. The effect is to expend the kinetic energy possessed at N in doing work against this component; that

MECHANICS OF SOLIDS. 65

is, to convert it into potential energy at E of the same amount as previously possessed at B. Hence, it will oscillate to and fro between B and E for an indefinite length of time.

98. Terms Defined. — A *Single Vibration* is the motion of the particle from B to E, whereas a *Complete Vibration* is the motion from B to E and back again. *The Time* or *Period of Vibration* is that consumed in moving from B to E and is half that of a *Complete Period*, which is that consumed in moving from B to E and back again, or more generally, in passing from any point to its next passage through that point in the same direction. Half the angle described by the swinging pendulum is the *Amplitude;* for example, BAN.

99. Laws of the Pendulum. — It is shown in Kinetics that if the amplitude of vibration does not exceed about three degrees, the time of vibration depends solely on the acceleration due to gravity and on the length of the pendulum, and is given by the formula $t = \pi \sqrt{\dfrac{l}{g}}$, in which $\pi = 3.1416$, $l =$ length, and $g =$ the acceleration. As this formula contains but two quantities subject to change of value, namely, length and acceleration, the following laws are evidently comprehended in it:

I. *The time of vibration is independent of amplitude, if the amplitude be small.*

II. *The time of vibration at any one place varies as the square root of the length.*

III. *The time of vibration varies inversely as the square root of the acceleration of gravity.*

IV. *The vibrations at any place are made in equal times.*

100. Experimental Verification.—The above laws may be partly tested as follows:

Suspend three lead balls as shown in Fig. 33. Let the lengths of the supporting threads, taken in each case to the centre of the ball, be 1 m., ¼ m., and ⅛ m. respectively. Find the time of a single vibration in each case by counting the number made in, say, 10 sec. These times will be found to be 1 sec., ½ sec., and ¼ sec. nearly, showing that the times are directly proportional to the square root of the lengths. Again, construct a pendulum, using an iron ball. Place a powerful magnet beneath it, so that the swinging ball will just escape touching it. If we determine the time of vibration, and then redetermine it with the magnet in position, we shall find that the magnet's attraction perceptibly lessens it. Placing the magnet beneath the iron ball is equivalent to moving the pendulum to a place where the acceleration due to gravity is greater. The experiment shows that the time would be lessened by such a change in the conditions.

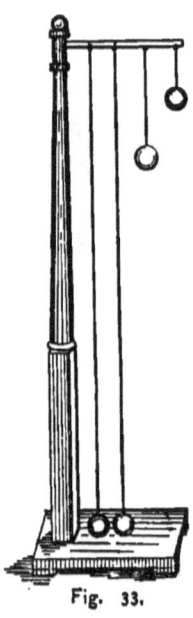

Fig. 33.

101. A Seconds Pendulum at any given place is one which makes a single vibration in one second. Its length for the latitude of New York can be computed by making $t = 1$ and $g = 980.19$ cm. in the formula $t = \pi \sqrt{\dfrac{l}{g}}$ and solving for l. [To be done by the student.] Since g increases on going from the equator toward either-pole, it is evident that l must increase.

102. A Compound Pendulum is a body supported on a horizontal axis, and free to swing about that axis under the action of the force of gravity. It is evident that every actual pendulum is compound. The most common

MECHANICS OF SOLIDS.

form consists of a heavy metallic weight (Fig. 34) called the *Bob*, supported by a slender rod, thin and flexible at the top.

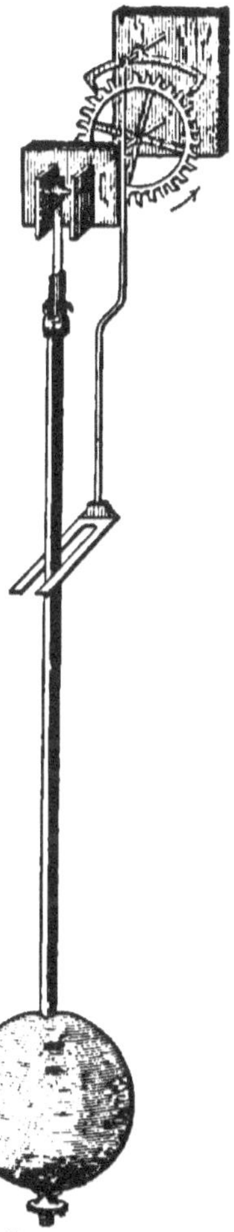

103. Centres of Suspension and of Oscillation. — Let AB (Fig. 35) represent a bar suspended from C, so as to have freedom of motion round that point; then C is the *Centre of Suspension*. Let G represent the centre of mass. This bar is a compound pendulum, and may be considered as composed of as many simple pendulums as there are material particles in it. According to the law of length, those particles nearer to C will strive to complete a vibration in less time than those more remote. The effect will be that the particles near C will accelerate the movement of those particles farther away, and those farther away will in turn retard the motion of those nearer to C. It is evident that there is some intermediate point, as D, which is neither accelerated by those above it, nor retarded by those below it, for the reason that its distance from C is such that, under the law of length, its time is that of the bar under consideration. This point is the *Centre of Oscillation*. As the particle at this point fulfils all the conditions of a simple pendulum whose time is that of the compound pendulum.

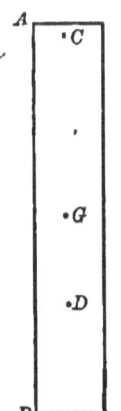
Fig. 35.

Fig. 34.

the length of the compound pendulum may be regarded as the distance between the centres of suspension and oscillation.

104. These Points Interchangeable. — Huyghens, a celebrated Dutch physicist, discovered that the centres of suspension and oscillation are interchangeable, that is, if a pendulum be reversed in position and be suspended from what was its centre of oscillation, the time of vibration will be unchanged. This discovery makes it possible to determine by experiment the true length of a pendulum, by observing the time of vibration of a compound pendulum, and then finding by trial the position of a second point on the other side of the centre of mass, about which the time of vibration is the same. The distance between these points is the length of the equivalent ideal pendulum.

105. Utility of the Pendulum. — Galileo is said to have been the first to point out that the pendulum performs its successive vibrations in equal intervals of time. This property commended it to him as a valuable measurer of time. The common clock is merely an instrument for keeping the pendulum in vibration, and recording the number of its vibrations. Since the vibration-period is affected by any change in the length of the pendulum, it is evident that if regularity is to be secured, invariability of length must be maintained. It is well known that heat lengthens rods and cold contracts them; hence it follows that clocks will lose time on exposure to heat, and gain on exposure to cold. To correct for such variations, the bob can be raised or lowered by means of an adjusting screw, thus changing the length of the pendulum. In astronomical clocks, the pendulums are made of two different metals, so combined that their changes in length act against each other, producing nearly perfect compensation.

MECHANICS OF SOLIDS.

The pendulum is of great value in measuring the acceleration due to gravity. This is evident from the formula $t = \pi\sqrt{\frac{l}{g}}$; for, if we determine t for a pendulum of known length at any place, we can readily compute the value of g.

Like all rotating or vibrating bodies, the pendulum maintains its plane of vibration unchanged. Foucault was the first to apply this fact in the experimental demonstration of the earth's rotation on its axis.

EXERCISES.

1. A toy-car whose mass is ½ lb. runs at the rate of 5 miles an hour on a level circular railway 20 feet in circumference; calculate the horizontal pressure on the rails in pounds, also in poundals.

2. A mass of 20 pounds is revolving uniformly, once in 5 seconds, in a circle whose radius is 3 feet. Find the centrifugal force.

3. A mass of 1 lb. is placed on the rim of a wheel 2 ft. in diameter, which revolves on its axis, and is otherwise balanced. The linear velocity of the rim being 30 ft. per sec., what is the pull on the axis caused by the mass of 1 lb.?

4. If a tower were 200 ft. high, with what velocity would a stone dropped from the top strike the ground?

5. A rifle ball is shot vertically upward with a velocity of 1,500 ft. per second; in what time will it reach the ground?

6. A stone thrown over a tree reaches the ground in 3 sec.; what is the height of the tree?

7. The acceleration due to gravity at Paris is 9.81 metres per second; what is the length of the seconds pendulum at that place?

8. What must be the value of g in order that a pendulum one metre long shall vibrate seconds?

9. What must be the length of a pendulum that shall make 70 vibrations per minute, when $g = 32.16$ ft.?

10. A pendulum 10 ft. in length makes 10 *complete* vibrations in 35 seconds; what is the value of g at the place?

VIII. SIMPLE MACHINES.

106. A Machine is any contrivance having for its object the transference or the transformation of energy. An electric lighting plant illustrates both of these uses. The energy of the steam is communicated to the moving parts of the engine, to be transferred by belts to the armature of the dynamo, where it is transformed into electric energy.

In mechanics it is customary to restrict the term "machine" to such devices as merely transfer energy. This is clearly the duty of those known as *Simple Machines* or *Mechanical Powers*. They are six in number, the *Lever*, *Wheel and Axle*, *Pulley*, *Inclined Plane*, *Wedge*, and *Screw*. All other machines are but combinations of two or more of these.

107. Forces Involved. — In every machine there are two forces involved, the *Weight* and the *Effort*. By the former is meant the resistance to be overcome, which may be a weight to be lifted in opposition to gravity, or a body to be moved in opposition to some force, as cohesion; by the latter is to be understood the force necessary to overcome such resistances.

108. Mechanical Advantage. — The general problem in machines consists in finding the ratio of the weight to the effort, in terms of some easily determined dimensions of parts of the machine in question. This ratio is called the *Mechanical Advantage*. In elementary discussions of machines it is the practice to neglect all friction and to assume that the parts are perfectly rigid and without weight.

109. Law of Machines. — Every machine must conform in its use to the principle of Conservation of Energy

(71), that is, *the work done by the effort is equal to the work done in overcoming the resistance.* Let the effort be represented by P, the weight by W, the distance through which the effort acts by D, and that through which the resistance is moved by D'; then $P \times D = W \times D'$; that is, *the effort multiplied by the distance through which it acts is equal to the load multiplied by the distance through which it is moved.*

110. The Efficiency of a Machine. — In all machines there are, in practice, two classes of resistances, — the *Useful*, that which the machine was specially designed to overcome, and the *Wasteful*, such as friction,[1] rigidity of cords, etc. Hence, in every machine two kinds of work are done, the *Useful* and the *Wasteful*, corresponding in character to these two kinds of resistances. The *Efficiency* of a machine is the ratio of the useful work done by the machine to the total work done on the machine. If we represent the *Efficiency* by M and the *Useful Work* by W_u, then $M = \dfrac{W_u}{P \times D}$. If this ratio were *unity*, the machine would be a perfect one, wasting no energy. If it were set running without doing useful work, it would continue to run forever, a condition which is clearly impossible.

111. The Lever is an inflexible bar movable about a fixed axis called the *Fulcrum.* The parts of the lever into which the fulcrum divides it are called the *Arms.*

---

[1] Friction is the resistance met with when one of two bodies in contact is made to move over the other. Its cause is found in the fact that the surfaces are not smooth; even when highly polished there are still minute irregularities. Hence, when brought in contact, some of the projections of the one drop into the depressions of the other, and energy is spent either in breaking off the projections or in lifting the body out of the depressions. Friction is, therefore, not a primary force, but is rather an absorption of energy by its transformation into some other form.

When the arms are in the same straight line the lever is *Straight*, otherwise it is *Bent*. If the fulcrum be between the points of application of the effort and the weight, the lever is of the *First Kind* (Fig. 36); if the weight be between the effort and the fulcrum, the lever is of the *Second Kind;* if the effort be between the weight and the fulcrum, the lever is of the *Third Kind*.

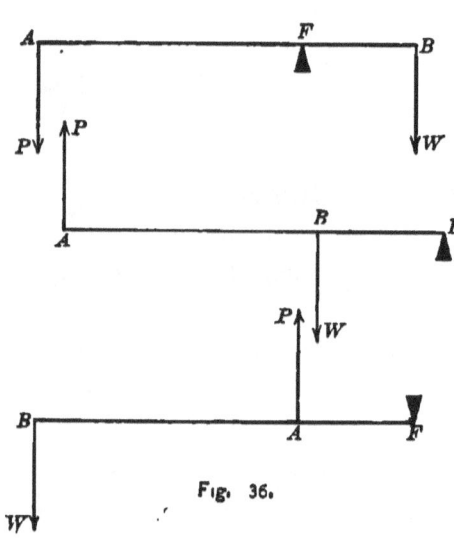

Fig. 36.

112. Illustrations. — The *Beam Balance* (Fig. 37) and the *Common Steel-yard* (Fig. 38) are levers of the first kind, differing from each other in that the former usually has equal arms. Scissors and most pincers are double levers of the *first kind*, the arm of the weight or resistance varying with the position of article being cut or held. The crow-bar when used for lifting a weight by resting one end on the ground is a lever of the *second kind*. Two persons carrying a weight slung from a pole are em-

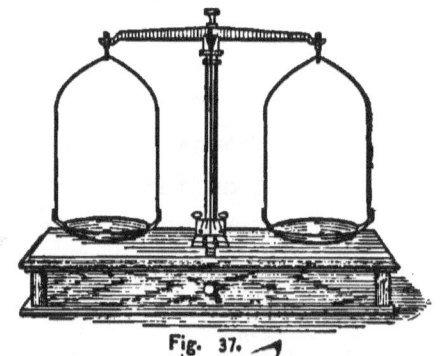

Fig. 37.

ploying a lever of the *second kind*. Table nut-crackers are double levers of this kind. In a swinging door we have a lever of either the *second* or the *third* kind according as the force which moves it is applied near the knob or near the hinges. (Where is the point of application of the weight?) When a weight is lifted in the hand, the forearm acts as a lever of the *third* kind, for the fulcrum is at the elbow and the effort is applied through the tension of the tendons at a point between the weight and the fulcrum.

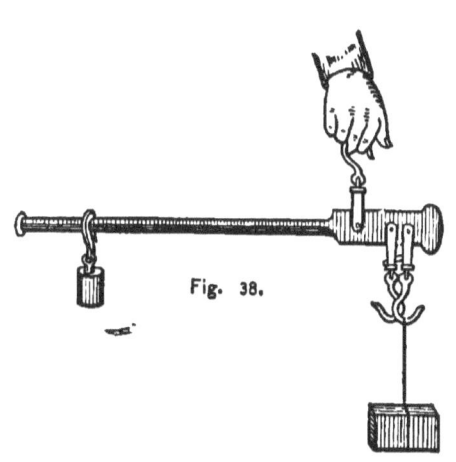

Fig. 38.

113. Mechanical Advantage of the Lever. — Let AB (Fig. 39) be a lever whose arms are FA and FB. For simplicity, suppose the bar to be without weight, or else loaded at one end so that on removing P and W it will balance about F. Tilt the lever, giving it the position of aFb, then P moves vertically through aC and W through bD. Therefore, by Art. 109, $P \times aC = W \times$

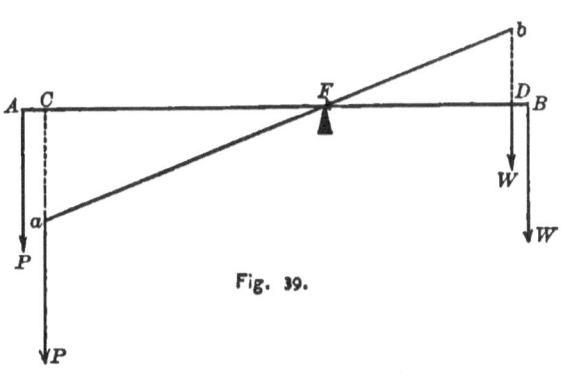

Fig. 39.

bD, or $\dfrac{W}{P} = \dfrac{aC}{bD}$. But by Geometry, $\dfrac{aC}{bD} = \dfrac{Fa}{Fb} = \dfrac{FA}{FB}$. Therefore $\dfrac{W}{P} = \dfrac{FA}{FB}$. Hence the following law:

The mechanical advantage of a lever equals the inverse ratio of its arms.

114. The Wheel and Axle (Fig. 40) consists of a wheel and cylinder fitted together so as to turn on the same axis. The effort is applied to the circumference of the wheel, B, and the weight to that of the cylinder, A, by means of a rope winding round it in a direction opposite to the motion of the wheel.

Fig. 40.

115. Illustrations. — A common form of the wheel and axle is the *Windlass* (Fig. 41), a machine used for raising weights, as earth from a well, brick to the top of a wall, etc. The effort is applied to the end of a lever, CF, called the *Crank* or *Winch*, which

Fig. 41.

may be considered as the radius of the wheel.

The Capstan (Fig. 42) illustrates the axle in a vertical position, the effort being applied by means of levers inserted in holes in the upper end. The machine is used in moving buildings, and on shipboard for raising the anchor or drawing the ship to the dock.

Fig. 42.

116. Mechanical Advantage of Wheel and Axle. — An examination of the machine shows that during each revolution of the wheel the effort acts through C, the circumference of the wheel, and the weight through c, the circumference of the axle. Therefore, by Art. 109, $P \times C = W \times c$, or $\dfrac{W}{P} = \dfrac{C}{c}$. By Geometry, $\dfrac{C}{c} = \dfrac{R}{r}$, R and r representing radii of the wheel and axle respectively. Therefore, $\dfrac{W}{P} = \dfrac{R}{r}$. Hence, the law:

The mechanical advantage of the wheel and axle equals the ratio of the radii of the wheel and the axle.

117. Modifications of the Wheel and Axle. — It frequently happens that the wheel and the axle are on separate axes, the motion of one being communicated to the other by means of friction, belts, or teeth on the circumferences of the wheels. In such cases the law is obtained by using, instead of the ratio of the radii, the ratio of the circumferences, or that of the number of teeth on these circumferences.

118. The Pulley is a wheel free to turn about an axis; the effort and the weight are attached to a cord which works in a groove cut in the circumference of the wheel.

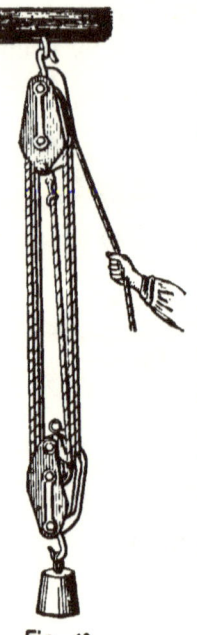

Fig. 43.

The frame supporting the pulley is called the *Block*, and the pulley is either *Fixed* or *Movable*, according as the block is stationary, or moves during the action of the effort.

119. Systems of Pulleys. — Combinations of pulleys may be made in endless variety, the principal ones being: 1st, when the same cord passes round all the pulleys; and 2d, when a separate cord is used for each pulley. In practice two or more pulleys are frequently mounted in the same block and turn on the same axis. An example of this is found in the common "Block and Tackle" (Fig. 43), used in moving buildings and in raising weights.

120. Mechanical Advantage of the Pulley. — The mechanical principle involved in all calculations with the pulley is the constancy of the tension in all parts of the cord. The only purpose served by the pulley itself is to diminish the friction. When the cord passes in succession around each pulley, as in Fig.

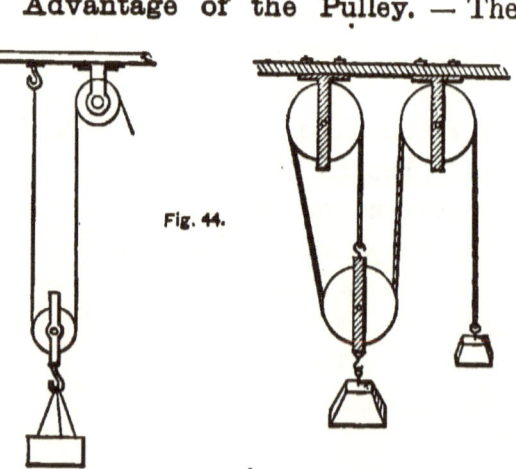

Fig. 44.

44, it is evident that the weight is sustained by the several parts of the cord, the tension on each part being P, the force applied. Hence, if there are n parts to the cord, the total tension going to support W is $P \times n$; that is, $W = P \times n$ and $\dfrac{W}{P} = n$. Therefore:

The mechanical advantage of the pulley when a single cord is used equals the number of parts of the cord which support the weight.

121. **The Inclined Plane** is a frictionless surface which is inclined to a horizontal surface.

When a body is placed on such a surface, the action of gravity is not only to produce pressure on the plane, but also to cause motion down the plane (57). To maintain the position of the body on the plane, force may be applied in one of three ways: 1st, Parallel to the face of the plane; 2d, Parallel to the base of the plane; 3d, Making an angle with the face of the plane other than the angle of the plane.

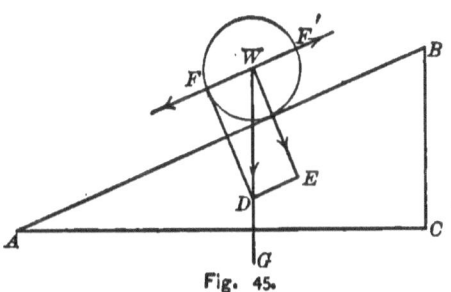

Fig. 45.

122. **Mechanical Advantage of the Inclined Plane.** — Case I. When the force is applied parallel to the surface of the plane, the value of the force acting down the plane can be found by resolving W into two components (Fig. 45), WE and WF, the former perpendicular to the plane and representing the pressure on the plane, the latter parallel to the plane and representing the magnitude of the force acting down the plane. WF must be balanced by an equal

force, WF', in order to maintain the body on the plane, and consequently $WF = P$. An inspection of the figure shows that the triangles DFW and ACB are similar. Hence, by Geometry, $\dfrac{WD}{WF} = \dfrac{AB}{BC}$. Since $WD = W$, $WF = P$, $AB = l$, and $BC = h$, then $\dfrac{W}{P} = \dfrac{l}{h}$, that is:

The mechanical advantage of the inclined plane when the force is applied parallel to the plane equals the ratio of the length of the plane to the height.

Case II. When the force is applied parallel to the base of the plane, the value of the force necessary to retain the body on the plane can be found by resolving W into two components, WE and WF (Fig. 46), the former perpendicular to the plane and the latter parallel to the base. WE is neutralized by the reaction of the plane, and WF by P acting in the opposite direction to it. Hence $WF = P$. By Geometry, the triangles DFW and ACB are similar, and $\dfrac{WD}{WF} = \dfrac{AC}{BC}$, or $\dfrac{W}{P} = \dfrac{b}{h}$, that is:

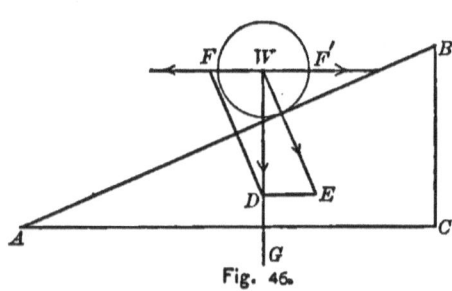

Fig. 46.

The mechanical advantage of the inclined plane when the force is applied parallel to the base of the plane equals the ratio of the base of the plane to the height.

Case III. When the force acts at an angle to the surface of the plane, the mechanical advantage cannot be obtained without the aid of Trigonometry. An approximate

solution can be obtained by a graphic process. For example,
a plane is inclined to the horizon at an angle of 30°; find the force necessary to sustain a weight of 100 lbs. on this plane, when applied at an angle of 60° to the horizon. Draw AB (Fig. 47), making an angle of 30° with AC; WD, representing

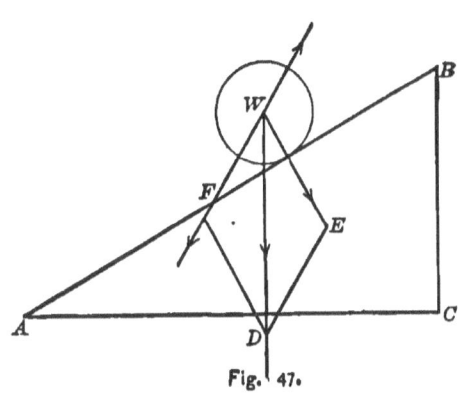

Fig. 47.

100 lbs., one inch long; and WF making an angle of 60° with AC. Resolve WD into the components WF and WE, the latter being perpendicular to AB. The value of WF will be the force required. By the aid of dividers and scale it is found to be .58 in. long. Hence $F = 58$ lbs. nearly.

123. The Wedge (Fig. 48) is merely an inclined plane, with the effort applied parallel to the base, causing the plane to move. The effort is generally applied by per-

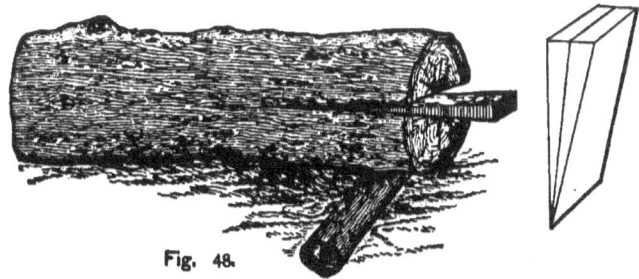

Fig. 48.

cussion, and although the principle of the machine is that of the second case of the inclined plane, yet no accurate determination of the mechanical advantage is possible, owing chiefly to the fact that it is used merely to force surfaces apart, and these act as levers materially aiding the effort applied. Cutting instruments, nails, pins, etc., are examples of wedges.

124. The Screw is a cylinder on the outer surface of which there is a uniform spiral projection called the

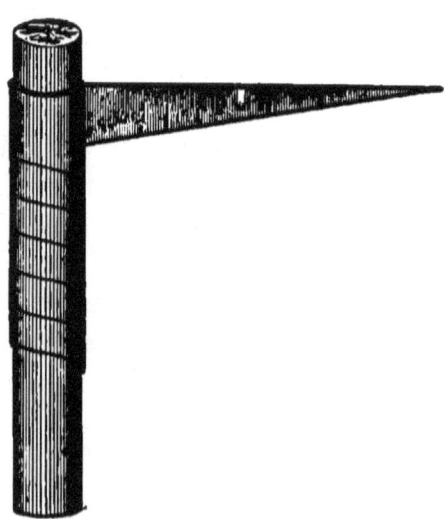

Fig. 49.

Thread. The faces of this thread are of the nature of inclined planes, as will be seen by wrapping a triangular piece of paper around a cylinder (Fig. 49). The hypothenuse of the triangle will trace out a spiral on the cylinder resembling the threads of a screw. The screw, when in use, works in a block called the *Nut*, on the inner surface of which is cut a groove which is the exact counterpart of the thread (Fig. 50). The effort is applied at the end of a lever fitted either to the nut or to the cylinder (Fig. 51). When the lever makes a complete revolution the screw or nut moves through a distance equal to that between two contiguous threads, called the *Pitch.*

125. Mechanical Advantage of the Screw. — As already stated, the screw is merely an inclined plane, and it is not a difficult matter to find the value of $\frac{W}{P}$

by applying the principle of the inclined plane. Since, however, the screw is usually combined with the lever, it will be quite as simple to find the ratio $\frac{W}{P}$ by applying the principle of work as expressed in the general law of machines (109). If the pitch be represented by d, and the lever arm by l, then when P makes a complete revolution, the work done will be $P \times 2\pi l$. Since the screw moves through the distance between two contiguous threads,

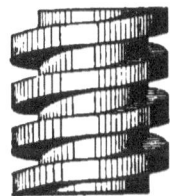

Fig. 50.

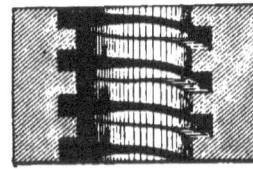

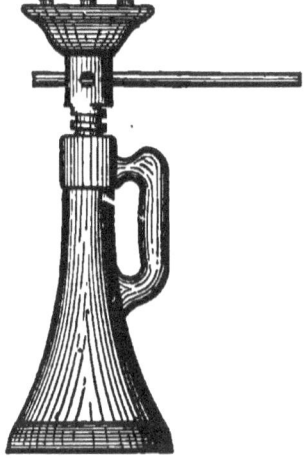

Fig. 51.

or d, while the arm is making one revolution, then the work done by the screw is $W \times d$. Hence $W \times d = P \times 2\pi l$, and $\frac{W}{P} = \frac{2\pi l}{d}$, that is:

The mechanical advantage of the screw equals the ratio of the distance traversed by the effort in one revolution to the pitch of the screw.

126. Applications. — The screw is used for overcoming great resistances, as in raising buildings, compressing hay or cotton, pulling stumps, propelling ships, etc. In the majority of cases it depends on friction for its efficiency, and hence it is not feasible to verify experimentally the law governing its action.

where great accuracy is required. An example of this is seen in the *Wire Micrometer* (Fig. 52), where a fine screw has its large circular head divided into a number of equal

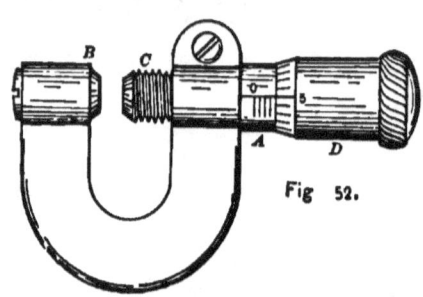

Fig 52.

parts, so that it can be turned through any required part of a complete revolution. If, for example, the pitch of the screw is $\frac{1}{50}$ of an inch, and the head is divided into 20 equal parts, then at each revolution of the screw the end advances or recedes $\frac{1}{50}$ of an inch; and if the head be turned through one of the divisions upon it, the end of the screw moves but $\frac{1}{20}$ of $\frac{1}{50}$ or $\frac{1}{1000}$ of an inch.

EXERCISES.

1. A man weighing 75 kilogrammes is lowered into a well by means of a windlass, the arm and axle of which are 80 cm. and 20 cm. in diameter. Find the force which must be applied to support him at any point during the descent.
2. Suppose that we have seven weightless pulleys, three movable and four fixed, connected by a single cord, and that a weight of 200 lbs. is raised; find the force applied.
3. A screw, the pitch of which is $\frac{1}{4}$ in., is turned by means of a lever 4 ft. long; find the force which will raise 250 lbs.
4. The thread of a screw makes 12 turns in a foot of its length; the effort is applied at the end of an arm 2 ft. long; it is found that when the effort is 30 lbs., it can just raise 1,200 lbs.; what portion of the effort is used in overcoming friction, and how many foot-pounds of work are done by the effort when the weight is raised 2 feet?
5. On a lever of the second kind a weight of 20 kilos. is suspended at a distance of 20 cm. from the fulcrum; the effort is 5 kilos.; find the length of the lever.
6. A weight of 1,000 lbs. rests on an inclined plane whose elevation is 1 ft. in 10; find the effort, applied parallel to the slope, necessary to keep it from moving down the plane.

CHAPTER III.

MECHANICS OF FLUIDS.

I. NATURE OF FLUIDS.

127. Fluidity. — Fluids do not possess invariability of form, as is seen in the readiness with which they assume the shape of the containing vessel. This is due to the ease with which the relative position of their molecules can be changed. *A Perfect Fluid* is one which offers no resistance to change of shape. Such a fluid is ideal, since every actual one, so far as known, offers resistance to the passage of a body through it; that is, it exhibits *Viscosity*. The degree of viscosity varies with the substance, being, for example, very large in tar, as compared with that of hydrogen gas. Some solids may not improperly be classed as fluids of large viscosity, since they slowly change their shape under the influence of pressure. Sealing-wax, pitch, and cobbler's-wax are familiar examples.

128. Liquids and Gases Compared. — Experiments show that liquids are slightly compressible, the compressibility varying with the nature and the temperature of the substance. Water, under a pressure of 14.7 pounds to the square inch, is reduced the 0.00005th part of the original volume, and mercury the 0.000005th part. On removing the pressure, in every case the liquid's original volume is exactly restored, showing the possession of perfect elasticity. Gases, on the other hand, are highly compressible and perfectly elastic. Under suitable conditions of temperature and pressure, any gas can be reduced to a liquid. MM. Pictet and Cailletet liquefied oxygen gas by subjecting it to a pressure of 300 atmospheres (144), at a temperature of — 29° C.

Exp. — Fill a small rubber foot-ball half full of air, and place it under a bell-jar on the air-pump table. As the air is exhausted from the bell-jar, the sides of the ball will be seen to move outward, and the ball will have the appearance of being fully inflated. What property of air is illustrated by this experiment?

Exp. — Push the piston into the cylinder of a pop-gun. If the piston be well fitted, on removing the hand it will return nearly to its original position. What properties of air are illustrated?

Is the fact that the piston does not return exactly to its original position inconsistent with the statement that gases possess perfect elasticity? Explain.

II. TRANSMISSION OF PRESSURE.

129. Pascal's Law. — Fig. 53 illustrates the position that a number of balls would assume on being placed in a vessel. If we endow these in imagination with freedom of motion and perfect elasticity, it is evident that on applying a pressure of, say, 10 grammes to the ball A, it will push the balls B and C apart, causing them to force apart F and K and D and E, just as if the pressure had been applied directly to each of these balls. These balls in turn act on those adjacent to

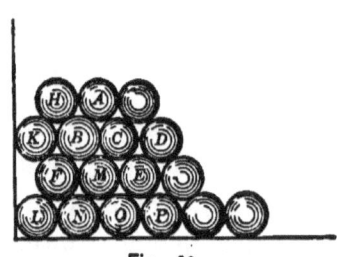

Fig. 53.

them, till the whole mass of balls is affected, each one as if the pressure had been applied directly to it. Therefore, the force, 10 grammes, applied to the ball *A*, becomes multiplied in its effect by as many balls as compose the mass, and accordingly the pressure exerted on the sides of the containing vessel is a number of grammes equal to ten times the number of balls touching them.

A body of fluid may be considered as made up of a number of molecules possessing perfect elasticity and freedom of motion, and hence if pressure be applied to any one or to a number of them, it will be transmitted in a manner similar to that described above. Pascal, a celebrated physicist of the seventeenth century, first stated the law of the transmission of fluid pressure, which is as follows:

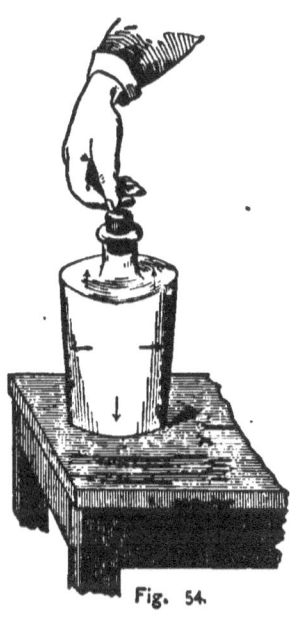

Fig. 54.

Pressure exerted on any given area of a fluid enclosed in a vessel is transmitted undiminished to every equal area of the interior of that vessel.

130. Illustrations. — **Exp.** — Fit accurately to the mouth of a thin-walled pint bottle a close-grained cork (Fig. 54). Fill the bottle full of water, and then force in the cork by pressure. The probable result will be a broken bottle. Explain.

Exp. — Glass-blowers make a form of syringe which is attached to a hollow sphere provided with several small openings (Fig. 55).

Fill the apparatus with water, and gently push the piston into the cylinder. The water will escape in a series of jets, and apparently the streams have equal velocities, notwithstanding the fact that only one of these jets is in line with the piston.

Exp. — An apparatus, shown in section in Fig. 56, can be easily made out of brass tubing and glass tubing. Fill it with water, and then insert the piston, holding the cylinder vertically. The water is forced up these several tubes to the same height, showing an equality of pressure in the different directions.

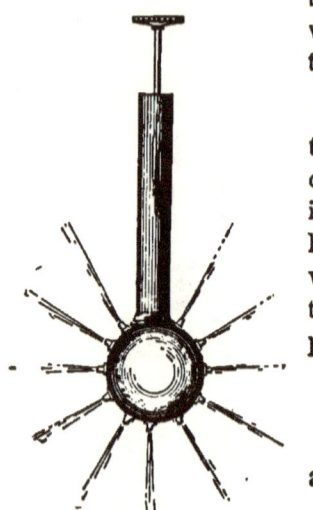

Fig. 55.

131. Hydraulic Press. — Probably the most convincing proof of the truth of Pascal's law is furnished by the *Hydraulic Press*. This instrument, shown in section in Fig. 57, consists of two cylinders, connected by the tube CCC. The small cylinder, aa, is provided with two valves at the bottom (not shown in the figure) similar to those in a force-pump (162), and with a small plunger worked by the lever L. Water is thus drawn from a surrounding reservoir and forced through the tube CCC into the large cylinder V, in which a large plunger, AA, works. The large plunger is thus forced upward and carries with it the plate BB' and the goods placed on it.

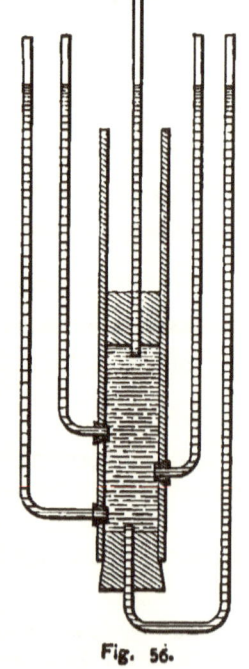

Fig. 56.

MECHANICS OF FLUIDS. 87

The force applied to the handle *L* is multiplied by means of the lever, producing a much greater force at *I*, pressing down the small plunger. This force is again multiplied by the ratio of the areas of the two plungers, a fact easily proved by a dynamometer. If, for instance, the

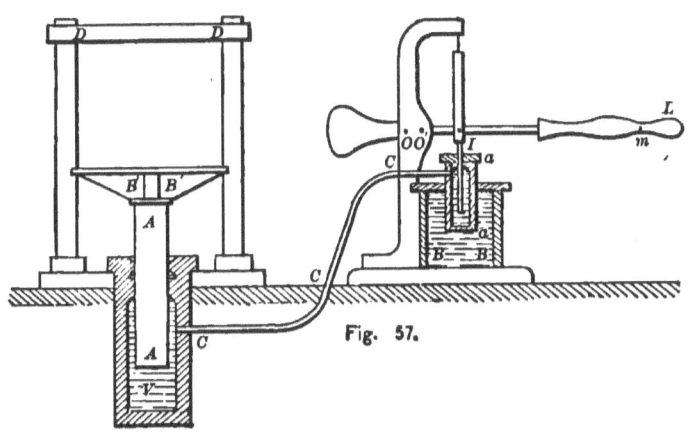

Fig. 57.

large plunger has 100 times the cross-sectional area of the small one, the multiplying power is 100.

III. PRESSURE DUE TO GRAVITY.

132. Pressure in Different Directions Compared. — Since the earth attracts each particle of a fluid toward its centre, each one must exert pressure on those beneath, and these will transmit it to all of the others, causing thereby lateral and upward pressures at each point. An examination of any fluid enclosed in a vessel will not reveal the existence of any currents, the only motion detectable being that of a species of molecular wandering, known as diffusion (37). This absence of currents would imply that these internal pressures at any point must be in equilibrium; that is, *the pressure exerted in a fluid by gravity at any point is equal in all directions.*

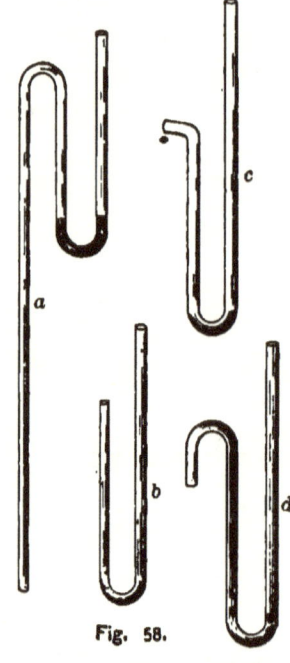

Fig. 58.

133. Proved for Liquids.—Exp. — Bend a stout glass tube into the form shown in Fig. 58 (*a*); and also three shorter pieces into the forms shown in (*b*), (*c*), and (*d*). In the U-shaped part of (*a*), pour enough mercury to rise about 1 cm. in each arm. Attach the tube (*b*) to the lower end of (*a*) with a short piece of rubber tubing, and hold the apparatus vertically in a vessel of water, observing the distance of the mouth of (*b*) below the surface of the water; and also the position of the mercury in (*a*). Now substitute successively the tubes (*c*) and (*d*) for (*b*), and submerge the apparatus to the same depth as before. The mercury in (*a*) will be found to register the same change of level in every case, proving that the pressure in a liquid at a point is equal in every direction.

134. Proved for Gases. — Exp. — Tie a piece of sheet-rubber over one end of a bladder-glass (Fig. 59), place it on the table of the air-pump, and exhaust the air. The rubber will be

Fig. 59.

pressed into the glass as if a weight were placed on it. If we turn the pump on its side, or even upside down, the result is the same. Hence, we see that the *air exerts pressure in every direction, downward, upward, and laterally.*

Exp.— Manufacturers of physical apparatus construct accurately fitting hollow metallic hemispheres, known as *Magdeburg Hemispheres* (Fig. 60), which can be connected to an air-pump, and the air exhausted. When this is done they will be held firmly together by the pressure of the atmosphere. In pulling them apart it matters not in what position we place them, the force necessary to effect the separation is the same for the same degree of exhaustion, showing that *the pressure exerted by the atmosphere at any point is equal in all directions.*

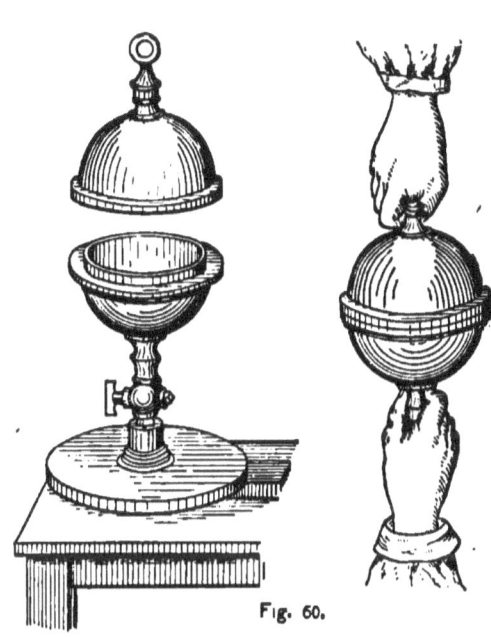

Fig. 60.

That the hemispheres are held together by the pressure of the atmosphere is shown by their falling apart when they are placed under a bell-jar on the air-pump table and the air around them is exhausted.

IV. PRESSURE ON THE BOTTOM OF A VESSEL.

135. Affected by Depth.—Exp.—Grind one end of a glass tube, 15 cm. long and 2.5 cm. in diameter, till a disk of brass or tin closes it water-tight (Fig. 61). Suspend the disk by a cord from one end of a scale-beam and counterpoise it. When the scale-beam is horizontal the disk should be in contact with the ground end of the vertically supported tube. Now place, say, 200 grammes on the scale-pan and pour water into the cylinder until its pressure detaches

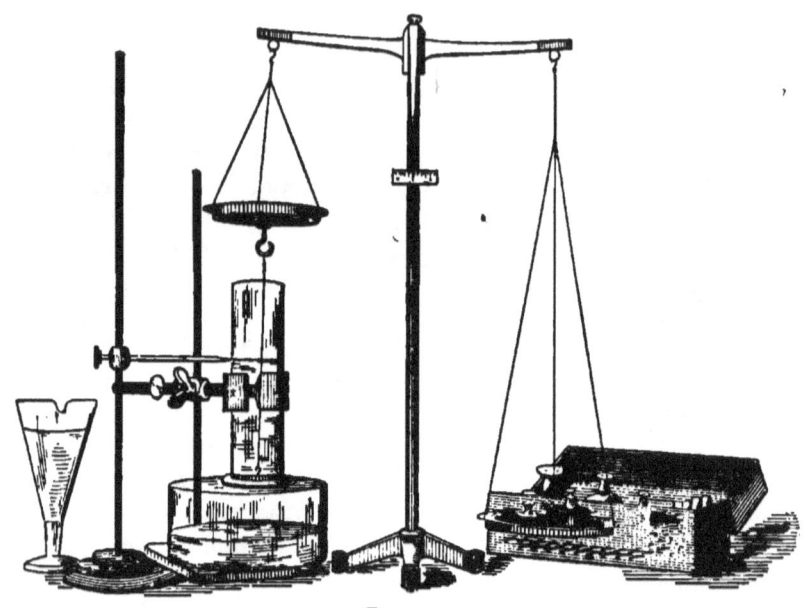

Fig. 61.

the disk, marking the depth. Repeating the experiment with double the weight on the pan, we find that the depth of water necessary to release the disk is twice as great, showing that

The downward pressure of a liquid is proportional to the depth of the liquid.

136. Affected by Kind of Liquid. — **Exp.** — Using the apparatus of the last experiment, we find that a certain depth of water is necessary to detach the disk. If, however, we use kerosene oil, a liquid specifically lighter than water, we discover that a greater depth is necessary, showing that

The downward pressure of a liquid depends on the density of the liquid (169).

137. Affected by Area of Base. — Exp. — As in Art. 135, prepare two glass cylinders, one having twice the cross-sectional area of the other. Proceeding as in that experiment, we find that the disk is detached from the larger cylinder when the depth of liquid is but half that required for the smaller one. Hence:

The downward pressure of a liquid is proportional to the area of the base.

Fig. 62.

138. Unaffected by Shape of Vessel. — Exp. — Proceed as in Art. 135, using successively three vessels differing in shape but having equal bases (Fig. 62). It will be found that with a given weight in the scale-pan, the disk will be detached when the depth of water is the same in each case. Hence:

The downward pressure of a liquid is independent of the shape of the vessel.

139. Rules for Pressure. — Since the downward pressure depends on the depth and nature of the liquid, and the area of the surface pressed upon, we derive the following rule:

The downward pressure of a liquid on a horizontal surface is equal to the weight of the column of liquid whose base is the area pressed upon, and whose altitude is the distance of the given surface below the surface of the liquid.

In applying this rule to water, the pressure in pounds can be obtained by observing that one cubic foot of water weighs 62.4 lbs.

Since the pressure at any point in a liquid is equal in all directions, it follows:

First. *The lateral pressure of a liquid on any vertical surface is equal to the weight of the column of liquid whose base is the area pressed upon, and whose altitude is the distance of the centre of area of this surface below the surface of the liquid.*

Second. *The upward pressure of a liquid on any horizontal surface is the same as the downward pressure on an equal surface similarly placed with respect to the surface of the liquid.*

V. EQUILIBRIUM IN FLUIDS.

140. Conditions of Equilibrium.— Any fluid remains at rest when the resultant of all the forces acting on it is zero. If for any cause a greater downward pressure is exerted at any point of the fluid than at others, it will not be balanced by the pressures transmitted from these parts, and hence motion of matter toward these parts is the consequence. This is seen in the case of currents in both air and water. The air or the water, as the case may be, for some cause or other may exert less pressure at one point than it does at points adjacent, and hence there is a movement of the fluid toward the point of less pressure.

141. Surface of a Liquid at Rest.— If a quantity of liquid is confined in a vessel, its surface to be at rest must be horizontal. For if it has any other form, as *ABD* (Fig. 63), any particle out of level, as *B*, will at once begin

MECHANICS OF FLUIDS. 93

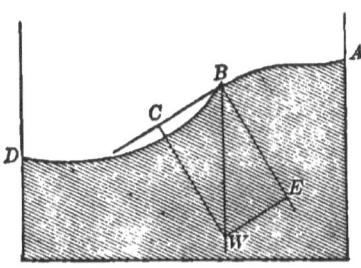

Fig. 63.

to move in obedience to the force BC, one of the components of the weight BW, of the particle. The other component, BE, being perpendicular to the surface of the liquid, is neutralized by the resistance of the liquid. When the surface is level BC vanishes and the motion ceases.

It is evident that when gravity is the only force acting on the liquid the surface is perpendicular to the plumb-line. In the case of a surface of large area, since all plumb-lines nearly meet at the centre of the earth, the surface is nearly spherical and would be exactly so were it not distorted by the centrifugal action due to the earth's rotation on its axis.

142. **Communicating Vessels.** — Fig. 64 illustrates a number of vessels of various shapes and capacities connected together through the horizontal tube, mn. On pouring water into one of them, it will rise evenly in the others, maintaining the same height in each. For if not, the column of greatest height would exert a downward pressure which would cause an equal upward pressure in each of the other tubes, and unless neutralized by an equal downward pressure in these tubes would cause the water

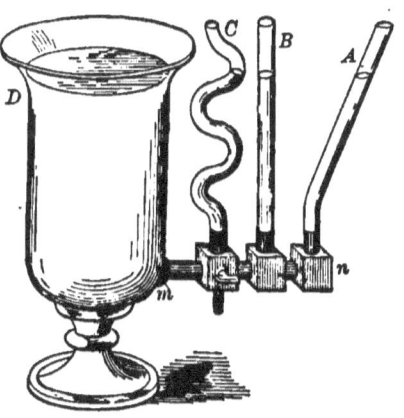

Fig. 64.

to move up these tubes till equilibrium is secured, that is, till the depth is the same as in the first one.

In systems of city water-works, we have an illustration of connected vessels on a large scale. Springs, fountains, water and spirit levels also illustrate the same principle.

VI. THE BAROMETER.

143. Torricelli's Experiment. — Select a stout glass tube, about 80 cm. long, and closed at one end. Fill it with mercury, close the open end with the finger, and invert it in a cup of mercury (Fig. 65). On removing the finger, the mercury settles away from the top end a few centimetres, leaving a vacuum. If we measure the height of this mercury column from time to time, we find that it varies. If we carry the apparatus up a high hill, we find that it settles away still farther. If we place it under a tall bell-jar on the air-pump table, the column also falls on removing the air. Hence, it is evident that atmospheric pressure sustains the mercury in the tube.

This experiment was first performed by Torricelli. For a full account of his work the student may consult Routledge's "History of Science."

Fig. 65.

144. Pressure Measured. — In the case of communicating vessels it was seen that the area of the cross-section did not affect the height at which the liquid stood, as in the case of Torricelli's experiment the size of the tube does not influence the height, provided it is not so small as to be affected by capillarity. Hence, if a tube of one square centimetre cross-section be employed, the height will be just the same as if a smaller, or a larger tube be used. Experiments made at the level of the sea show an average height of 76 cm. for the mercurial column. Then, with a tube of one square centimetre cross-section, the volume of mercury supported in it will be 76 ccm. Since 1 ccm. of mercury weighs 13.596 grammes, the weight of the mercury will be 76 times 13.596 grammes, or 1033.3 grammes; that is, the atmosphere exerts an average pressure on each square centimetre of surface at the level of the sea of 1033.3 grammes, or 14.7 pounds per square inch. Such a pressure is called an *Atmosphere*.

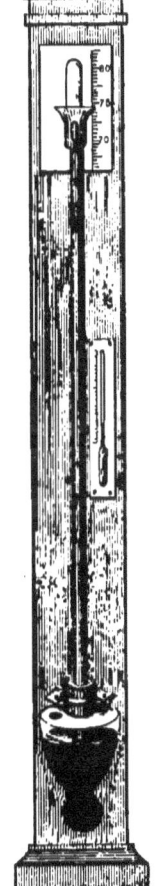

Fig. 66.

145. The Mercurial Barometer, in its simplest form, consists of a Torricellian tube about 86 cm. (nearly 34 in.) long, firmly attached to an upright support (Fig. 66). A scale, whose zero is at the surface of the mercury in the cistern, is fastened by the side of the tube, to give the height of the mercury column. All readings must be taken with the tube in a vertical position. Why? With the fluctuations in the height of the mercury in the tube, it is evident that the surface of the mercury in the cistern changes its level, and consequently affects the zero of the scale. To obviate this error, the top of the cistern is made large, or the bottom is adjustable.

Since mercury is a very heavy liquid, the changes in the barometric readings are seldom very great, and small changes in atmospheric pressure may escape notice. Glycerine is about one-tenth as heavy as mercury. If it were used as a substitute, a tube nearly ten times as long as that used in mercurial barometers would be required. Such an instrument would be very sensitive, since a change of one millimetre in the reading of the mercurial barometer would be about one centimetre on the scale of the glycerine barometer.

146. **Barometric Variations.** — Since the mercury in the tube of the barometer is sustained by the pressure of the column of air resting on it, any change in this pressure will produce a change in the barometric reading. Experiments show that changes of this kind are going on continually at every place. Certain very slight changes are found to be periodical, but the greater changes follow no known laws. These irregular movements point to corresponding fluctuations in the pressure of the air, and, consequently, herald important atmospheric movements.

147. **Uses of the Barometer.** — The barometer is a faithful indicator of all changes in atmospheric pressure. Experience has shown that these changes are generally indicative of conditions which determine changes in the weather according to the following rules:

I. *The rising of the mercury indicates the approach of*

MECHANICS OF FLUIDS. 97

fair weather; the falling of the mercury shows the approach of foul weather.

II. *A sudden fall of the mercury denotes the approach of a storm.*

III. *A rise of the mercury frequently precedes the clearing up of a storm.*

Since the pressure of the atmosphere diminishes as we ascend from the surface of the earth, the height of the barometric column is evidently connected with the altitude of the place. Various rules have been proposed for determining the altitude of a place, based on a knowledge of the barometric reading at that place, but they are more or less inexact. See Cooke's " Chemical Physics," Art. 167.

EXERCISES.

1. The diameter of the cylinder and the tube of a hydraulic press are, respectively, 4 inches and 1 inch, the lever 5 feet long, and the piston-rod 10 inches from the fulcrum; what resistance can be overcome by a force of 100 pounds?

2. What is the upward pressure on an inch board 12 feet long and 12 inches wide, sunk 6 feet under water?

3. What pressure resists the separation of a pair of Magdeburg hemispheres 4½ inches in diameter when the barometric reading is 29 inches?

4. If we let bubbles of air pass up through a glass of water, why will they become larger as they ascend?

5. In digging canals, ought the curvature of the earth to be taken into consideration?

6. On applying the pressure-gauge to a water-pipe in a building it registers 75 lbs. How high is the supplying reservoir above that point?

7. What would be the height of a water barometer when a mercurial one reads 28.5 inches?

8. Calculate in dynes per square centimetre the atmospheric pressure when the barometer stands at 78 centimetres.

9. Find the pressure in grammes per square centimetre at the bottom of a vessel 1.75 metres deep, full of mercury.

10. A canal lock is 12 feet wide, 8 feet deep, and 50 feet long. Find the total pressure on the gate and on the bottom of the lock.

VII. THE AIR-PUMP AND CONDENSER.

148. The Air-pump, as its name denotes, is a device for removing air or any gas from a vessel, and depends for its action on the elastic force of the gas. Fig. 67 repre-

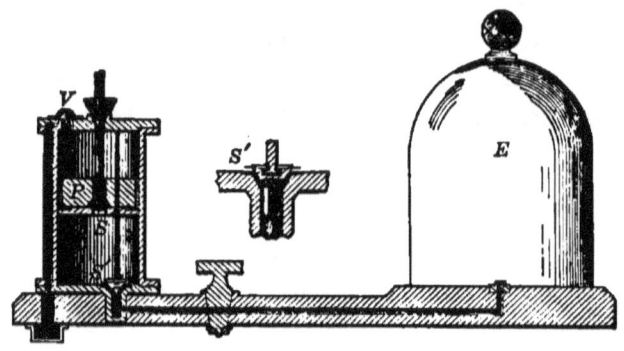

Fig. 67.

sents the essential parts of one of the best forms made at the present time. A piston P, with a valve S in it, works in a cylindrical barrel, communicating with the outer air by a valve at its upper end, shown in the figure at V, and with the receiver E by the horizontal tube. The valve S'' is carried by a rod which passes through the piston, fitting tightly enough to be lifted by the piston when the up-stroke begins; but its ascent is almost immediately arrested by a stop near the upper end of the rod, and the piston slides on this rod during the remainder of the up-stroke. This allows the air from E to flow into the space below the piston. In the down-stroke, the piston first carries down

MECHANICS OF FLUIDS. 99

the valve rod with it and closes the valve S''. During the same movement the valve S is mechanically opened, and the enclosed air passes through it into the upper part of the cylinder. The ascent of the piston again closes S; and as soon as the air is sufficiently compressed, it opens the valve V and escapes. Each complete double stroke of the piston removes a cylinder full of air, but as it grows rarer with each double stroke, the mass removed each time grows less. On account of the tendency of gas to fill the containing vessel, irrespective of quantity, the removal of all the air from the reservoir is not possible, although we could continually approach a vacuum were it not for the unavoidable mechanical defects of the pump, such as leakage of valves, and the like.

149. Experiments with Air-pump. — A number of experiments requiring the use of the air-pump have already been given. To these are added the following:

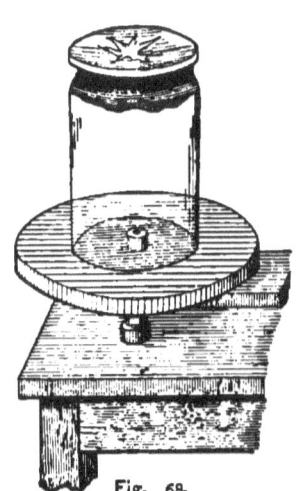

Fig. 68.

1. *The Bladder-glass Experiment.* — Over the end of a cylindrical receiver cement a disk of writing paper (Fig. 68). When dry, place it on the table of the air-pump and exhaust the air. The paper will break with a loud report. Explain.

2. *The Bacchus Experiment.* — Select two four-ounce bottles, fit to one of them a perforated stopper. Connect the two bottles by a bent tube reaching nearly to the bottom of each (Fig. 69). Fill the stoppered one nearly full of water and place them on the table of the air-pump beneath the receiver. Exhaust the air. Explain why the water

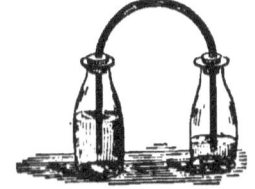

Fig. 69.

runs out of the stoppered bottle and runs back on admitting air into the receiver.

3. *The Vacuum Fountain.* — Fit a good perforated stopper to a large bottle (Fig. 70), and through the stopper insert a jet-tube. Connect the bottle to air-pump with a rubber tube and exhaust the air. Disconnect from the pump and insert the tube in a vessel of water. Why does the water rush into the bottle? How long a bottle would be necessary to prevent the jet from striking the bottom of the bottle?

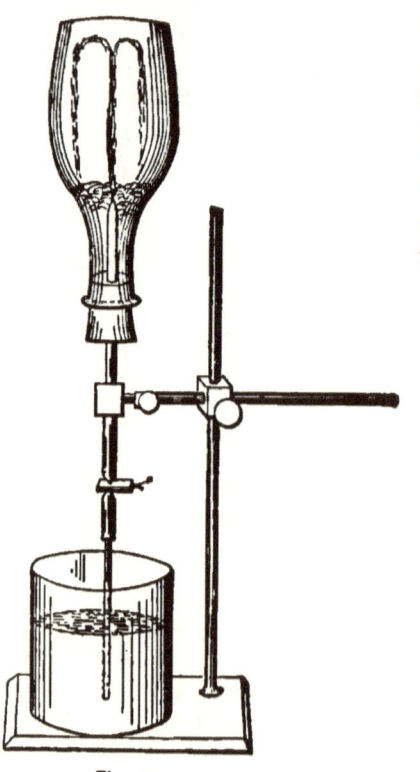

Fig. 70.

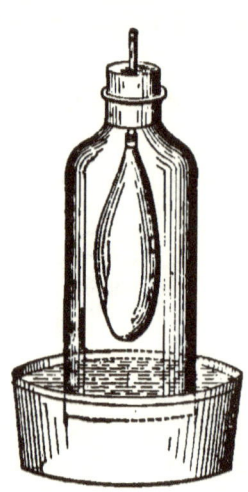

Fig. 71.

4. *The Lungs Experiment.* — Tie a glass tube about 20 cm. long into the mouth of a large rubber toy balloon. Place the balloon within an open-topped receiver with the tube passing through a cork accurately closing the top of the receiver (Fig. 71). Hold the receiver in a deep vessel of water and account for the enlargement of the balloon on lifting the receiver, and its contraction on pushing the receiver downward.

This experiment illustrates the mode of action of the lungs in breathing, the surface of the water in the receiver being the dia-

MECHANICS OF FLUIDS.

phragm. A more marked action can be got by placing the receiver on the table of the air-pump instead of in the vessel of water.

5. *Liquid Condition of Matter dependent on Pressure.* — Fill a test-tube with water and invert it in a tumbler of water. Introduce into the tube a few drops of sulphuric ether. The ether will rise to the top of the tube. Place the apparatus under the air-pump receiver and exhaust the air. The water will fall in the tube (why?) and the liquid ether will vaporize and fill the tube. On admitting the air to the receiver it resumes its original condition.

150. The Condensing Pump. — If the air-discharging pipe of an air-pump be connected with a suitable vessel, the pump during its action will force air into it. Such a device would be a *Condensing Pump.* Since the valves of the ordinary air-pump will not stand high pressures, a pump designed as in Fig. 72 is better adapted to forcing gas into vessels. The plunger is solid, and in the bottom of the cylinder are two valves, one opening inward and the other outward. As the piston moves upward the gas is admitted through the left-hand tube, the valve being lifted by the pressure of the gas below it. When the piston descends this valve closes and the right-hand one opens, affording an exit for the confined gas.

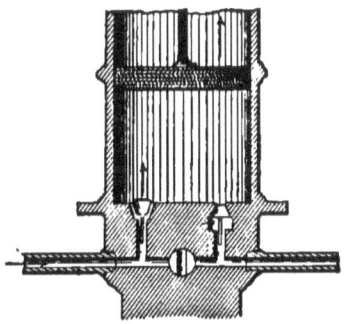

Fig. 72.

This machine is evidently an air-pump when the left-hand tube is connected to a receiver, but is not capable of producing a very low degree of exhaustion, owing to the heavy valves necessary to give it strength for compressing gases.

151. Applications. — Both the air-pump and the condenser are extensively used in the arts. Sugar-refiners employ the air-pump to reduce the boiling point of the syrup; manufacturers of soda-water employ the condenser to charge water with carbonic acid gas; in pneumatic dispatch tubes, now employed in many large cities for rapidly transferring small packages, both instruments are used, the former to remove the air from the iron tubes in front of the closely fitting carriage, and the latter to force compressed air into the other end of the tube, thereby propelling the piston-like carriage with great velocity. The condensing pump is also employed to improve the draught of furnaces, to facilitate the ventilation of mines, and to operate machinery in places difficult of access.

VIII. LAW OF BOYLE.

152. Law of Boyle. — It was shown experimentally in 1662 by Robert Boyle, an English physicist, that the following simple relation exists between the amount of compressibility of a gas and the pressure applied:

The volume of a gas varies inversely as the force of compression, the temperature remaining constant.

If the volume V of a gas under a pressure p becomes V' on changing the pressure to p', then $\frac{V}{V'} = \frac{p'}{p}$. Hence $Vp = V'p'$, that is, *the product of a given volume of gas by its pressure is constant for the same temperature.*

153. The Law Verified. — Exp. — *First,* bend a stout glass tube to the shape of the letter J, the short arm being about 25 cm. long and closed at the end, while the long one is about 170 cm.

MECHANICS OF FLUIDS. 103

long and terminates in a small funnel. Fasten the tube to a wooden support, and place a scale of equal parts by the side of each arm with the zero just above the bend of the tube (Fig. 73). Pour into the tube sufficient mercury to fill it exactly to the zero of the scale. The air enclosed in the short arm will be under a pressure indicated by the barometer-reading at the time of the experiment. On pouring more mercury into the long arm, the air column in the short arm will contract, since it is under a pressure greater than at the beginning by that produced by a column of mercury whose length is equal to the difference between the two mercury columns. If a number of volume-readings and corresponding pressure-readings be obtained, it will be found that the product of the corresponding ones is nearly constant, showing that when the pressure exceeds that of the atmosphere, *the volume of a gas varies inversely as the pressure.*

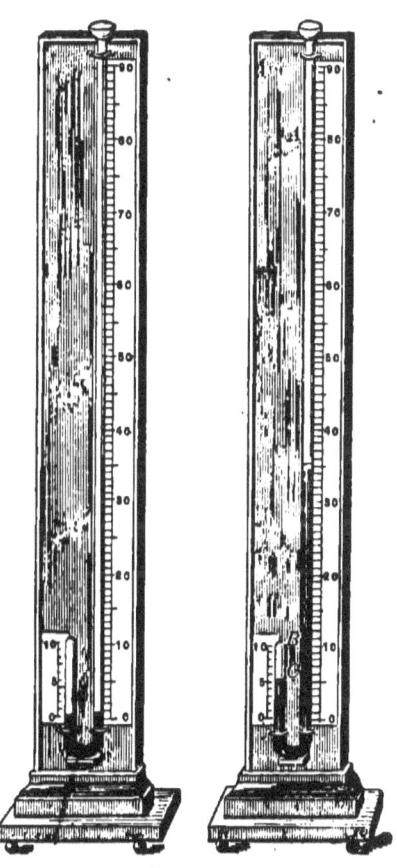

Fig. 73.

Secondly, close one end of a stout glass tube that is about 80 cm. long. Fill it with mercury to within, say, 20 cm. of the end. This 20 cm. of air is under a pressure indicated by the barometric reading at the time. Close the tube with the finger and invert it in a vessel of mercury. The mercury will fall part way in the tube, and the air column will occupy considerably more than 20 cm. in length. It is evident that the column of mercury exerts pressure in opposition to the air that presses on the mercury in the vessel. Hence, the air in the tube is under a pressure represented by the difference between the barometric reading and the mercury column in the tube. It will be found that if the initial volume of air be multiplied by the barometric reading, the product will nearly equal that of the final volume multiplied by the reduced pressure, showing that for a pressure less than that of the atmosphere, *the volume of a gas varies inversely as the pressure.*

154. Inexactness of the Law. — Experiments made by Dulong, Arago, and others show that the law of Boyle is only approximately true, some gases being more compressible than others, and all of them deviating from the law as they approach the point of liquefaction.

IX. THE SIPHON AND THE WATER-PUMP.

155. The Siphon, in its simplest form, is a U-shaped tube designed for conveying fluids from one vessel over an elevation to another vessel at a lower level by means of atmospheric pressure. To set it in action, the usual way is to fill the tube with the liquid, close the ends, place the shorter branch in the vessel of liquid, and open the ends. The flow will continue as long as the liquids in the two vessels are at different levels, and the shorter branch dips into the liquid. The

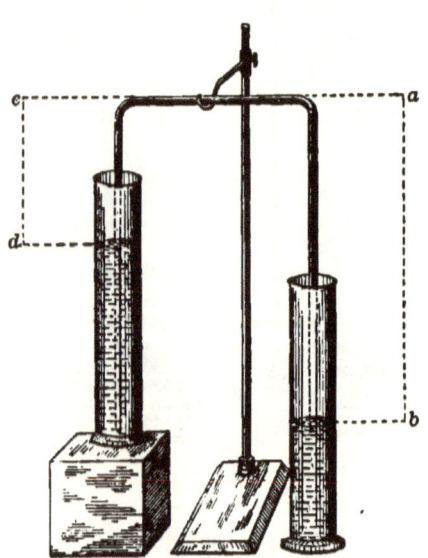

Fig. 74.

MECHANICS OF FLUIDS. 105

siphon may also be started by suction; in the case of corrosive liquids a suction tube (Fig. 75) is attached in a manner to prevent contact of the liquid with the mouth. The vertical distance of the highest part of the siphon above the surface of the liquid in the vessel being emptied equals the length of the *Short Arm* of the siphon, as *cd* (Fig. 74); the vertical distance of this highest point above the outlet of the siphon, if not within the liquid of the receiving vessel, equals that of the *Long Arm*, as *ab*. When the outlet is within the liquid, the measurement must be made to the plane of the surface of the liquid and not to the end of the tube.

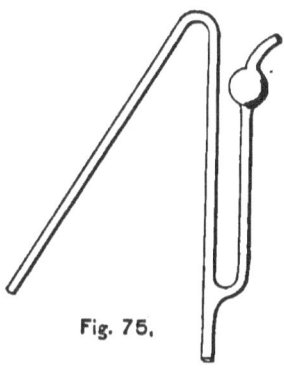

Fig. 75.

156. The Siphon's Action dependent on the Inequality of the Arms. — **Exp.** — With the long arm of a siphon connect a piece of rubber tubing, so that the length of that arm can be varied by raising or lowering the end of this tube. Set the siphon in operation and you will find that the rate of flow will increase as you lengthen the outer arm, and decrease as you shorten it, becoming nothing when the arms are equal in length.

Exp. — Construct a siphon of the form shown in Fig. 76, where the short arm is provided with a jet-tube opening within a bottle. On increasing the length of the long arm the force of the fountain jet within the bottle will be found to increase.

These experiments show that the arms of the siphon must be unequal, and that the velocity of the flow depends on the excess of the long arm over the short one.

157. Its Action dependent on Atmospheric Pressure. — **Exp.** — Using a siphon of small bore, set it in action on the air-pump table. Cover it with a bell-jar, and as quickly as possible exhaust the air. The liquid will stop flowing, but will resume flowing on admitting the air.

Exp.—Use the apparatus of the last experiment with mercury as the liquid. On exhausting the air it will be found that when the

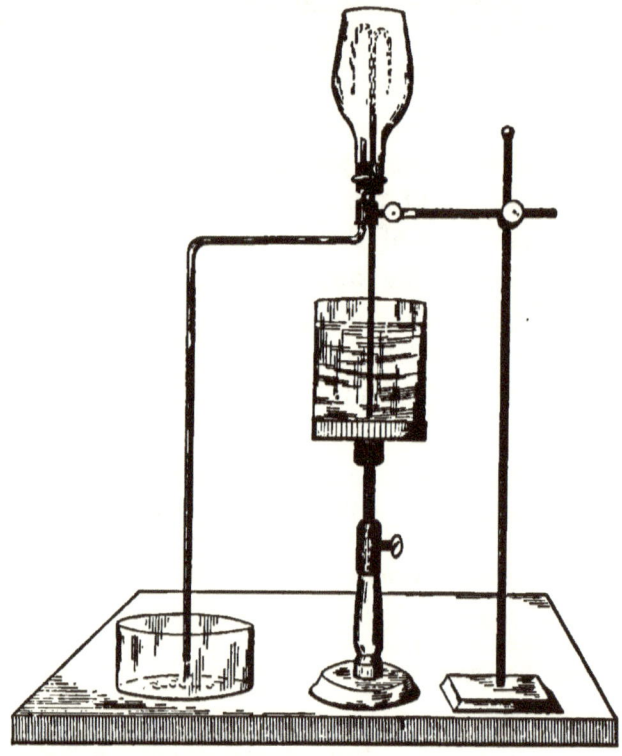

Fig. 76.

pump-gauge shows that the elastic tension of the air in the receiver measured on the gauge about equals[1] the difference of the arms of the siphon, the flow stops. Hence we see that the atmospheric pressure is necessary to the working of the siphon, and that the elevation over which a liquid can be siphoned is equal to the

---

[1] On account of the cohesion of the mercury the column does not break till the pressure-gauge registers less than the height of the siphon.

MECHANICS OF FLUIDS. 107

height of a column of that liquid which the atmospheric pressure will support.

158. The Siphon Explained. — The foregoing experiments justify the following explanation of the siphon: The atmospheric pressure at d (Fig. 74), by Pascal's Law,

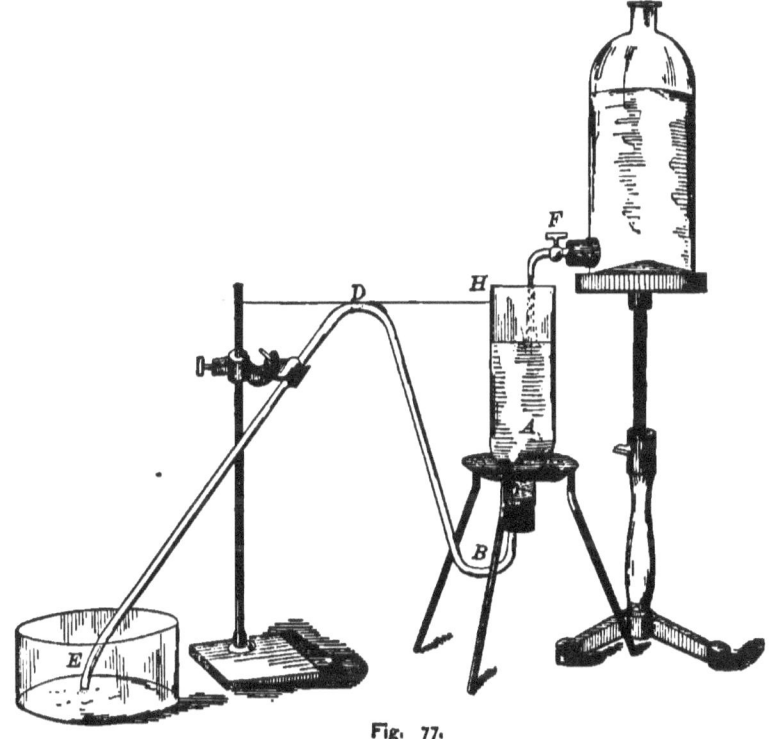

Fig. 77.

exerts an equal upward pressure in the tube; let this pressure be represented by p. This is lessened by the downward pressure w of the liquid in that arm. Hence there is a resultant force acting upward in the tube equal to $p - w$. Similarly there is a force acting upward in the long arm equal to $p - w'$, in which w' is the downward pressure of the liquid in that arm. Now, $p - w$ and $p - w'$

are forces acting in opposite directions, $p-w$ being the greater. Hence, their resultant is $(p-w)-(p-w')$ or $w'-w$; that is, there is an unbalanced force acting toward b, measured by a column of liquid whose depth is the difference between the arms, and causing the liquid to move toward b, a vacuum in the tube being prevented by the atmospheric pressure at d. It follows from this discussion that the elevation over which a liquid can be siphoned can not exceed the height of a column of that liquid which the atmospheric pressure will support.

159. The Intermittent Siphon. — Exp. — Cut off the bottom of a litre bottle, and support it in the ring of the iron stand. Bend a glass tube into the form shown in Fig. 77, and fit it to the bottle. The outlet E should be below the inner face of the cork, and the point D should be below the top of the vessel A. Adjust the tube F to supply water at a rate somewhat less than the siphon can carry off. As the vessel A fills with water the air is driven out of the siphon; and when the point H is reached the water passes over into the arm DE, thereby starting the siphon. The flow will continue till the vessel is emptied, and then will stop till the vessel is refilled to H, when the action will be renewed.

Fig. 78.

Natural intermittent springs and fountains, except geysers, are explained on the theory that a reservoir slowly fed by drainage or by a small stream has a siphon-shaped outlet (Fig. 78) acting in a manner exactly like the above intermittent siphon.

160. Applications. — The siphon is used for conveying water over small hills. (What is the height of the highest hill over which it is possible to siphon water?) It is also of service when either the top or the bottom of a liquid is to be drawn off without disturbing the other part.

161. The Suction Pump. — In the suction pump, a piston c, in which there is a valve opening upward, moves practically air-tight in a cylinder, at the lower part of which is an aperture fitted with a valve v, also opening upward (Fig. 79). From this aperture a pipe s leads down to the water in the well, the end of the pipe being below the surface of the water. When the piston is drawn upward, the valve in it is closed by the pressure of the air above it, and a vacuum will be formed in the cylinder below it. The pressure of the air in the pipe s will open the valve v, and the space between the piston and the water will be filled with air under a reduced tension. Hence, the pressure of the air on the water in the well will force the water up the tube

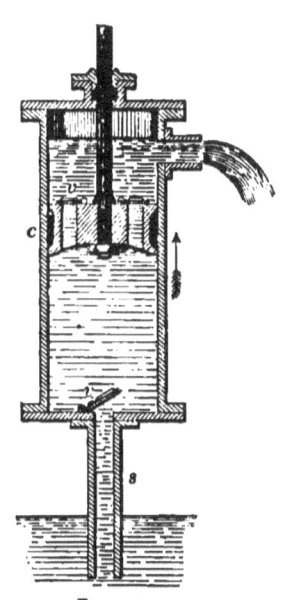

Fig. 79.

s to a height sufficient to restore equilibrium. When the piston descends, the valve v will close, while the valve in the piston will open and allow the air in the space below the piston to escape.

Thus it appears that each double stroke of the piston removes some air from the cylinder, thereby lessening the tension, and causing the water to rise in the pipe higher and higher from the pressure of the air on the water in the well. If *v* is distant from the water in the well less than 34 feet, provided the barometric reading is 30 inches, the water will be forced through *v* into the space above the piston, and by it will be lifted till it flows out of the aperture on the side.

162. The Force Pump. — In this pump (Fig. 80) the piston is solid, and the aperture through which the water escapes is placed between it and the valve *v*, and is closed by a valve opening outward from the cylinder. The explanation of the working of the pump is similar to that given for the suction pump. As in the suction pump, the valve *v* at the top of the pipe *s* must be within 34 feet of the water in the well. The height to which the water can be forced in the pipe *d* depends on the force applied to the piston.

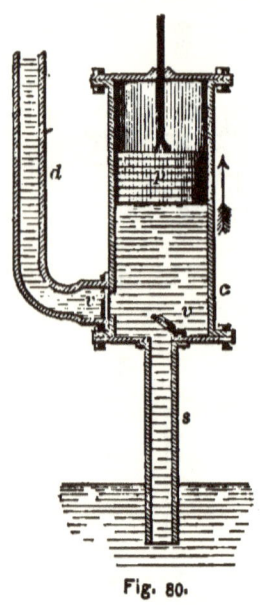
Fig. 80.

In powerful pumps, the water usually passes into an air-chamber called the *Air-Dome*. Its object is to give steadiness of flow to the water from the delivery-tube. Fire-engines and almost all pumps operated by steam are constructed in this way.

X. THE PRINCIPLE OF ARCHIMEDES.

163. The Buoyant Force. — Exp. — Suspend a stone from the hook of a spring-balance, and mark the position of the index. Now submerge the stone in a vessel of water; the index-reading is

MECHANICS OF FLUIDS. 111

less. If we use salt water the loss is still greater; but if we use kerosene it is not so great.

Exp. — In the pans of a short-beam hand-balance, counterpoise a large cork with a piece of lead. Fill a large, wide-mouthed jar with carbon dioxide, and support the balance in it. The equilibrium will be destroyed, the cork acting as if it were held up by some agent. If a jar of hydrogen be used instead, or if the air be exhausted from the jar, the effect is exactly the opposite.

It appears from these experiments that when a body is immersed in a fluid it loses in weight, or rather it is partially supported by the fluid, the amount of support varying with the nature of the fluid. This upward force is called the *Buoyant Force*.

164. Cause of Buoyant Force. — If a body be submerged in a fluid, it will be subjected to pressure on all sides. Let a cube of marble be immersed in water, as shown in Fig. 81. The lateral and opposite faces, a and b, will be equally pressed in opposite directions. The same will be true of the other pair of lateral faces. On d there will be a downward pressure equal to the weight of the column of water resting on it. This column has the face d for a base, and dn for height. On c there will be an upward pressure equal to the weight of a column of water whose base is c, and whose depth is cn. The resultant of these opposing pressures will be a force equal to their difference and acting upward. This difference will be the weight of a column of water whose base is c, and

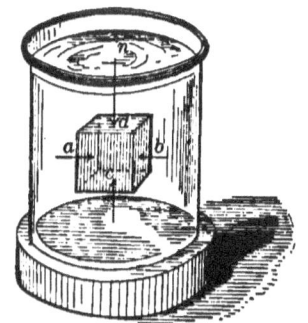

Fig. 81.

whose depth is *cd*; that is, the weight of the water displaced by the cube.

This truth, discovered by Archimedes, may be enunciated as follows:

A body submerged in a fluid is buoyed up by a force equal to the weight of the fluid displaced by it.

165. Experimental Proof. — Exp. — Procure a solid metallic cylinder 3.5 cm. long, and 1.9 cm. diameter. Its volume will be very nearly 10 ccm. Suspend it by a fine thread or hair beneath one of the pans of a balance, and counterpoise it. Now hold a dish of water beneath it, so that the cylinder is submerged. The equilibrium will be destroyed, and will be restored by placing a 10-gramme weight in the pan above the cylinder. Since the cylinder displaces 10 ccm. of water, weighing ten grammes, and loses ten grammes in weight when submerged, it follows that the body was buoyed up by a force equal to the weight of the liquid displaced.

166. True and Apparent Weight. — We have seen that when a body is submerged in a fluid, it is pressed upward by a force equal to the weight of the fluid displaced. Hence, every body in air is buoyed up by the weight of the air displaced by it, making its *apparent* weight differ from its *true* weight. [What correction must be applied to give true weight?]

167. Floating Bodies. — When a body is placed in a fluid, it may displace a weight of fluid *less than* its own weight, *equal to* its own weight, or *greater than* its own weight. In the first case, the upward pressure of the fluid will be less than the weight of the body and hence the body will sink. In the second case, the upward pressure will equal the weight of the body, and hence the body will be in equilibrium, remaining in the liquid wherever placed.

MECHANICS OF FLUIDS. 113

In the third case, the upward pressure will exceed the weight of the body, and hence the body will move upward till the portion remaining within the fluid displaces enough to cause an upward pressure equal to the weight of the body. Hence, the *Principle of Flotation:*

A body floats on a fluid when it displaces a volume of the fluid whose weight equals that of the body.

168. Principle of Flotation Proved. — Exp. — Cut a wooden bar 20 cm. long and 1.5 cm. square. Bore in one end a hole such as will contain shot sufficient to give the bar a vertical position when floating in water. Graduate a millimetre scale on one edge of the bar. Find the weight of the bar in grammes and then ascertain what part of it is submerged when floating in water. The volume of the submerged portion will be that of the water displaced; and allowing one cubic centimetre of water to weigh one gramme, we have the measure of the buoyant force. This will be found to be nearly equal to the weight of the bar, the difference being readily accounted for by the unavoidable errors of observation. Hence, *a floating body displaces its weight of the sustaining fluid.*

XI. DENSITY AND SPECIFIC GRAVITY.

169. The **Density** of a substance is its mass per unit of volume. In practice, it is convenient to consider it as the mass in grammes of a cubic centimetre of the substance at 0° C.

The Specific Gravity of a substance is the ratio of its mass to that of an equal volume of some standard. In practice, it is customary to consider it as the ratio of its mass, if the substance be either solid or liquid, to that of an equal volume of distilled water at 4° C.; but if gaseous, it is compared with either air or hydrogen at 0° C. and 76 cm. pressure.

170. To find the Density of a Solid insoluble in Water. — Find the weight of the body in air, and also its weight as it hangs suspended in ice-water. Let w grammes equal the weight in air, and w' its weight in water. Then, by the principle of Archimedes, $w - w'$ equals the weight of the water displaced. For most purposes one cubic centimetre of ice-water may be considered as weighing one gramme. Hence $w - w'$ will express in cubic centimetres the volume of the body, and $\dfrac{w}{w-w'}$ will equal the weight of one cubic centimetre, that is, $D = \dfrac{w}{w-w'}$.

171. To find the Density of a Solid soluble in Water. — Find the weight of the body in air, and also its weight as it hangs suspended in some liquid of known density in which it is not soluble. Let w equal the weight of the body in air, and w' the weight in a liquid whose density is s. Then, $w - w'$ equals the weight of a volume of the liquid equal to that of the body. Why? Since 1 ccm. of the liquid weighs s grammes, the volume of the liquid displaced, that is, the volume of the solid, must equal $\dfrac{w-w'}{s}$. Hence, $\dfrac{w-w'}{s}$ cubic centimetres of the solid weigh w grammes, and $D = w \div \dfrac{w-w'}{s} = \dfrac{ws}{w-w'}$.

172. To find the Density of a Solid lighter than Water. — Weigh the body in air. Then, to make it sink in water, attach to it a heavy body whose weight in water is known, and find the weight of the two bodies when submerged in ice-water. Let w equal the weight of the given body in air, a and b that of the sinker in air and in water respectively, and w' that of the body and sinker in water. Then $w + a$ equals the weight of both in air, $w + a - w'$ that of the water displaced (why?) and hence the volume in centimetres of the body and sinker combined, and $a - b$ the volume of the sinker. Why? Therefore, $w + a - w' - (a - b) = w - w' + b =$ the volume of the body, and $D = \dfrac{w}{w - w' + b}$.

173. **To find the Density of a Liquid.** — *First. Method with the Specific Gravity Bottle.* The specific gravity bottle (Fig. 82) is a bottle constructed to hold a given weight of distilled water at a specified temperature. Its capacity is usually 10, 25, 50, 100, or 1,000 grammes at 4° C. To use it, find the weight of the bottle, when empty, and also when filled with the given liquid. Let a be the weight of the empty bottle, w that of the bottle when filled with the liquid, and 100 grammes the amount of water required to fill the bottle. Then, $D = \dfrac{w - a}{100}$. Why?

Fig. 82.

Secondly. By weighing a Solid in it. Weigh a solid in air, then in ice-water, and finally in the liquid. Let w be the weight in air, w' in water, and w'' in the liquid. Then, $w - w'$ equals the weight of water displaced (why?), and hence the volume in centimetres of the solid and also that of the liquid it displaces; and $w - w''$ equals the weight of the liquid displaced by the solid. Why? Therefore $D = \dfrac{w - w''}{w - w'}$.

Thirdly. By the Hydrometer. The **Hydrometer** consists

116 ELEMENTS OF PHYSICS.

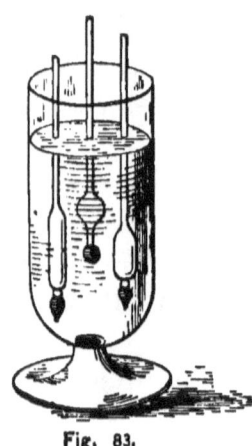

Fig. 83.

essentially of a straight stem, either wood, glass, or metal (Fig. 83), weighted at the lower end so as to float in a liquid in a vertical position. By observing the depths to which it sinks in two different liquids, the relative weights of these liquids can be determined. The stem is graduated either in some arbitrary manner or by trial, the zero being placed at the point to which it sinks in pure water. These instruments are used chiefly to determine the degree of concentration of such liquids, as acids, alcohol, milk, etc., and are graduated specially for the liquid under inspection.

174. To find the Specific Gravity of a Body. — An inspection of the preceding problems will show that in each case the weight of a volume of water equal to that of the body is obtained. If the weight in air be divided by this quantity, the quotient will be the specific gravity. It should be noticed that this weight of water when expressed in grammes is numerically equal to the volume of the body in cubic centimetres. Hence, the specific gravity of a body, when water is the standard, is numerically equal to the density in the metric system. If other units, as inch, foot, ounce, pound, be employed, this coincidence does not appear.

√ **175. Useful Formulæ.** — By definition $D = \dfrac{M}{V}$; hence $M = VD$ and $V = \dfrac{M}{D}$. Since the weight varies as the mass M, then we may write W instead of M. See

MECHANICS OF FLUIDS.

"Elementary Practical Physics," Vol. I., Chap. V., by Stewart and Gee. When English units are employed, $W = V \times 62.4$ lbs. $\times$ sp. gr. where V is cubic feet, and $V = \dfrac{W}{\text{sp. gr.} \times 62.4}$.

EXERCISES.

1. Why can stones be moved under water so much more easily than out of water?
2. A body weighs 62 grammes in air and 42 grammes in water; find its density.
3. A solid weighs 100 grammes in air and 64 grammes in a liquid of density 1.2; what is its density?
4. A specific gravity bottle, when filled with water, weighs 64.485 grammes, and, when filled with methylated spirit, 53.462 grammes. If the bottle weighs 15.063 grammes, what is the density of the liquid?
5. A solid weighs 120 grammes in air, is found to weigh 90 grammes in water, and 78 grammes in a strong solution of zinc sulphate; what is the density of the solution?
6. Find the density of a solid from the following data:
  Weight of solid in air . . . 0.5 g.
  " " sinker in water . . 3.5 g.
  " " solid and sinker in water, 3.375 g.
7. A bar of aluminium (density 2.6) weighs 54.8 grammes in air; what will be the loss of weight when it is weighed in water?
8. A body which weighs 24 grammes in air is found to weigh 20 grammes in water; what will be its apparent weight in alcohol of density 0.8?
9. The density of sea-water is 1.025; calculate the pressure in grammes per square centimetre at a depth of 40 metres below the surface of the sea.
10. A tube 120 cm. long holds 600 g. of mercury ($D = 13.6$); find its cross-section and internal diameter.
11. Compute the weight of a lead ball ($D = 11.3$) one inch in diameter.

12. Compute the diameter of an iron ball (D=7.8) whose weight is 300 grammes.

13. An ounce of silver (D=10.15) is suspended in water. Find the tension of the supporting string.

14. If the density of sea-water is 1.025, what portion of an iceberg (D=.917) is above water?

CHAPTER IV.

HEAT.

I. NATURE OF HEAT.

176. Heat, a Form of Energy. — Previous to about 1840 it was generally believed that *Heat* was an invisible and extremely subtle fluid without weight, which, by entering bodies and possibly combining with them, caused all thermal phenomena. The experiments of Davy, Rumford, Joule, and many others, have demonstrated the incorrectness of this view, and have led to the adoption of the modern *Kinetic Theory*. This theory, briefly stated, is as follows: The molecules of all bodies have a certain amount of independent motion, generally very irregular. When this molecular agitation is increased, the body is warmed, and when decreased it is cooled. Heat is the energy of this motion; it can be transferred from one body of matter to another, or can be transformed into other types of energy.

177. Temperature is the thermal condition of a body which determines the transfer of heat between it and any body in contact with it. If two bodies, A and B, are brought together, and if neither parts with any heat to the other, then they are at the same temperature. If A parts with some of its heat to B, A is at a higher temperature than B; and since B receives heat from *A*, it is at lower temperature. The temperature of a body depends on the kinetic energy of the molecules composing it, and is wholly independent of the quantity of matter. A pint of water in a vessel may be at a much higher temperature than the water in a lake, yet the latter contains a vastly greater quantity of heat, owing to the greater quantity of water, as would be seen if we should utilize it in melting snow.

178. Hot and Cold. — A body feels *hot* when it is imparting heat to our body, and *cold* when our body imparts heat to it. A cold body is at a lower temperature than our body, and, consequently, heat flows from our body to it; in the case of a hot body, the conditions are reversed.

179. Sensation Unreliable. — **Exp.** — Select three basins, *A*, *B*, and *C*. Fill *A* with hot water, *B* with cold water, and *C* with tepid water. Hold one hand in *A* and the other in *B* for a minute; then transfer both to *C*. The water of *C* will feel cold to the hand from *A* and warm to the hand from *B*.

Hence, it is evident that the sense of touch cannot be depended on to give accurate information regarding the relative temperature of bodies.

180. Measurement of Temperature. — **Exp.** — Fit to a short test-tube a perforated stopper, through which passes a glass tube, say 15 cm. long. Fill the apparatus with colored water part way up the glass tube, marking the height. Hold the test-tube for a few minutes in hot water; the volume of the liquid is noticeably increased.

Hence, we conclude that the heating of a body is attended by an increase of its volume. Experiments show that the amount of expansion of a body is definitely re-

HEAT.

lated to the increase of temperature. An application of this fact is to be found in the thermometer.

II. THE THERMOMETER.

181. A Thermometer is an instrument for measuring temperatures. The most common form consists of a narrow tube of uniform bore, having a bulb at one end, and hermetically sealed at the other (Fig. 84). This bulb and part of the tube are filled with some liquid, generally mercury. Either on the tube, or the supporting frame, is a scale of equal parts for measuring the rise or the fall of the liquid.

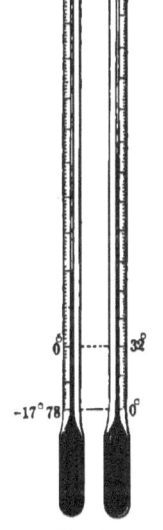

Fig. 84.

182. To Construct a Mercurial Thermometer. — A capillary glass tube of uniform bore is selected. A bulb, either spherical or cylindrical, is blown on one end. Part of the air is expelled by heating, and while in this condition the open end of the tube is dipped into a vessel of pure mercury. As the tube cools, mercury is forced into it by atmospheric pressure. Enough mercury should be introduced to fill the bulb and a small part of the tube at ordinary temperatures. Heat is again applied to the bulb till the expanded mercury fills the tube ; and while in this condition the top of the tube is sealed by fusing the glass in the blowpipe flame. If the work has been successfully done, the tube will contain no air, as may be easily ascertained by inverting it when cold and observing whether the mercury falls to the end on jarring.

183. Necessity of Fixed Points. — In order that the indications of different thermometers may be compared, some uniform plan of graduation is necessary. It is evident that spaces of equal lengths on two tubes may not represent equal changes of temperature, since one tube may be of finer bore than the other, and the bulbs may not be equal in volume. In consequence of either of these differences, equal changes in the height of the two mercurial columns will not indicate the same change of temperature. Why? Hence, two points must be found on every thermometer stem, each of which represents an invariable temperature. They are known as the *Fixed Points*, and correspond respectively to the temperature of *melting ice* and of *boiling water*. Both of them are found by experiment to be constant under the same conditions.

184. Location of Fixed Points. — The thermometer is packed in finely broken ice, as far up as the mercury extends. The containing vessel is provided with an opening at the bottom to let the water run away. After standing in the ice for several minutes the position of the mercury is marked on the tube. This is called the *Freezing-Point*.

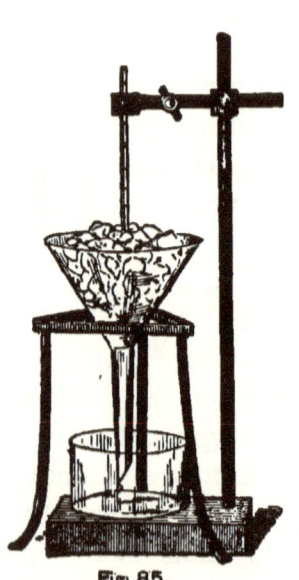

Fig. 85.

The *Boiling-Point* is found by observing the position of the mercurial column when the thermometer is wholly enveloped in steam under an atmospheric pressure of 76 centimetres (29.922

inches). Since this pressure is difficult to obtain, it is customary to observe the position of the column as the instrument is enveloped in steam and compute the position for a pressure of 76 cm. by the rule that a difference of

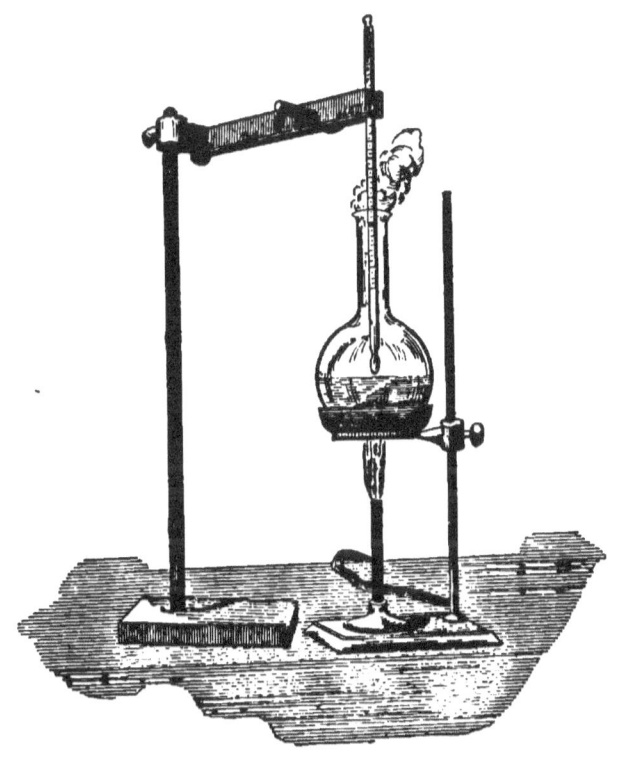

Fig. 86.

2.68 cm. in atmospheric pressure causes a difference of 1° C. in the temperature of the steam.

185. The Scale. — The distance between the fixed points is divided into equal parts called *Degrees*. The number of such equal parts being wholly arbitrary, different scales have come into use. In the *Fahrenheit Scale*

(Fig. 87), the freezing-point is marked 32° and the boiling-point 212°, the space between being divided into 180 equal parts; in the *Celsius* or *Centigrade*, the freezing-point is marked 0° and the boiling-point 100° the space between being divided into 100 equal parts; in the *Reaumur*, the freezing-point is marked 0° and the boiling-point 80°, the space between being divided into 80 equal parts. Each of these scales is extended beyond the fixed points as far as desired. The divisions below 0° are read as negative; for example — 10° signifies 10 degrees below zero. The scales are distinguished by affixing the initial letter of the name; for example, 5° F., 5° C., and 5° R. signify 5 degrees above zero on the Fahrenheit, Centigrade, and Reaumur scale respectively.

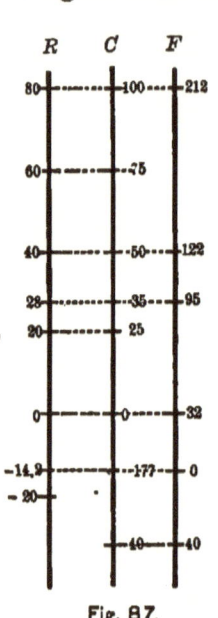

Fig. 87.

186. Relation of the Scales. — Since 180° F. = 100° C. = 80° R., then 1° F. = $\frac{5}{9}$ of 1° C. = $\frac{4}{9}$ of 1° R.; 1° C. = $\frac{9}{5}$ of 1° F. = $\frac{4}{5}$ of 1° R.; and 1° R. = $\frac{9}{4}$ of 1° F. = $\frac{5}{4}$ of 1° C. Hence, to convert Fahrenheit degrees into Centigrade degrees, multiply the number by $\frac{5}{9}$, and to convert Centigrade into Fahrenheit, multiply the number by $\frac{9}{5}$. (How would you change Centigrade into Reaumur or Fahrenheit, and conversely?) In converting any reading into the Fahrenheit scale, 32° must be added to the product, since the zero of that scale is 32° below the zero of the others. In the converse problem, 32 must be subtracted before multiplying. The following formulae will be found convenient: F. = $\frac{9}{5}$ C. + 32 = $\frac{9}{4}$ R. + 32; C. = $\frac{5}{9}$ (F. — 32) = $\frac{5}{4}$ R.; R. = $\frac{4}{9}$ (F. — 32) = $\frac{4}{5}$ C.

HEAT. 125

EXERCISES.

1. Express in the Centigrade scale the following: 30° F., — 4° F., 25° R., — 10° R., — 40° F.
2. Express in the Fahrenheit scale the following: 200° C., 20° C., — 4° C., — 15° R., — 40° C.
3. Express in the Reaumur scale the following: 75° C., 16° C., — 39° F., 212° F.
4. How would it affect the readings of a mercurial thermometer, if, after graduating, the bulb should contract?
5. How would it affect a thermometer if the bore grew larger from the neck of the bulb to the top end?

187. Limit to the Employment of the Mercurial Thermometer. — Since mercury freezes at — 38°.8 C., and boils at 350° C., it is evident that beyond these temperatures the mercurial thermometer cannot be employed, and even near these limits its indications are untrustworthy on account of changes in the rate of expansion of mercury, especially at the upper end of the scale. For temperatures below — 38° C., alcohol is substituted for mercury. This liquid freezes at about — 130° C. To estimate temperatures much above 300° C., the expansion of a metallic rod is employed, or an air thermometer with a porcelain or platinum bulb.

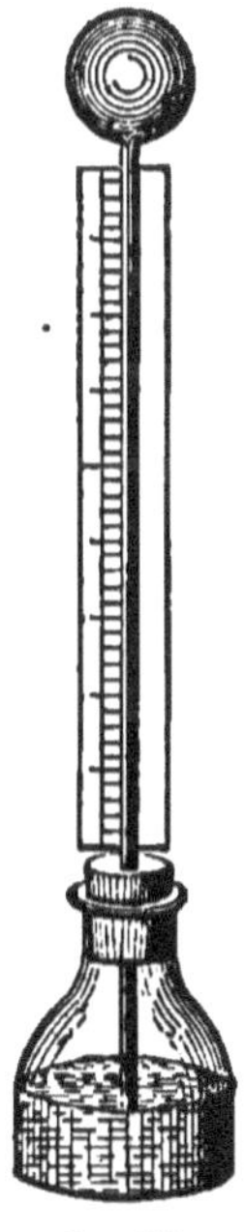

188. The Air Thermometer, in its simplest form, consists of a glass bulb attached to a tube of small bore, the lower end of which dips into colored water. The tube is supported in a vertical position, in front of a scale of equal parts. In Fig. 88 the cork of the bottle serves as a support. (Ought it to fit the bottle air-tight?) By

Fig. 88.

heating the bulb some of the air is expelled. On cooling, a column of colored water rises in the tube. Any slight change of temperature in the air of the bulb changes the height of the liquid column. When used to make comparisons with the mercurial thermometer it is a complex instrument, because corrections must be made for changes due to variations in the atmospheric pressure. Why? The instrument is remarkable for its sensitiveness, that is, for large movements of the index for small changes of temperature.

EXERCISES.

1. How would you construct a water thermometer?
2. If the bulb of a thermometer be plunged into hot water, the mercury at first falls. Why?
3. What effect will it have on the distance between the fixed points to use a tube with a very small bore? To use a large bore?
4. Why is a cylindrical bulb to be preferred to a spherical one for a thermometer?
5. What effect would the presence of air in the thermometer tube have on the indications of the instrument?

III. SOURCES OF HEAT.

189. The Sources of Heat may be divided into two great classes: those which are *Natural*, over which man has no control beyond that of utilizing them; and those which are *Artificial*, of man's own devising. Among the former, we may mention the *Sun* and the *Earth's Interior;* among the latter, are all those cases in which some existing form of energy is transformed into heat.

190. Heat due to Collision. — Exp. — Hold a piece of hardened steel between the thumb and forefinger of the right hand,

and strike a glancing blow against the sharp edge of a piece of flint. Sparks will be seen at each blow; their number and brilliancy will be found to depend upon the vigor of the blow struck.

Exp. — Pound a strip of lead with a hammer. Its temperature will be noticeably increased, as can be ascertained by touching it, or by bringing it in contact with an air thermometer. If the amount of pounding be increased, a higher temperature will be indicated.

These experiments are illustrations of the transformation of mechanical energy into heat. The falling hammer is endowed with energy, which is transferred in part to the lead on collision, thereby increasing the irregular vibratory movement of its molecules, which manifests itself as an increase of temperature. The same explanation applies to the heat produced by striking the flint with steel.

191. Heat due to Compression. — **Exp.** — Place a small piece of tinder in the cavity at the end of the piston of a Fire-Syringe (Fig. 89). Introduce the piston into the barrel and force it in quickly. Withdraw it with as little delay as possible; the tinder will probably be ignited.

The work done on the piston accounts for the increased heat energy of the inclosed air.

192. Heat due to Friction. — Savages kindle fire by rapidly twirling a dry stick, one end of which rests in a notch cut in a second dry piece. The axles of carriages and the bearings in machinery are heated to a high temperature when not properly lubricated. The heating of drills and bits in boring, the heating of saws in cutting timber, the burning of the hands by a rope slipping rapidly through them, the

Fig. 89.

stream of sparks flying from an emery wheel, are illustrations of the transformation of part of the energy due to their motion into heat, or energy of the molecule.

193. Heat due to Chemical Action. — **Exp.** — Place a small piece of phosphorus[1] on a board and drop on it a few crystals of iodine. In a few seconds the phosphorus will burst into a blaze, owing to the heat developed by these substances combining to form a compound of a red color known in chemistry as the iodide of phosphorus.

This experiment is but one of the many which chemistry furnishes showing that in most cases where chemical combinations are effected there is an evolution of heat. The combustion of wood, coal, etc., are instances of chemical action, the oxygen of the air combining with the constituents of the substances. The potential energy due to the affinity between the substances combining is transformed into kinetic energy and appears in the heat generated.

194. Animal Heat is due to oxidation. During respiration oxygen enters the lungs, passes by osmosis through the walls of the cells of the lungs into the blood, and combines with the carbonaceous matter found there, converting it into carbon dioxide and water. This chemical action is attended by heat, the amount depending on the amount of oxidation. Any cause increasing the supply of oxygen will increase the amount of heat, and accordingly raise the temperature of the body.

IV. DISTRIBUTION OF HEAT.

195. Exp. — Hold one end of a metallic rod in a Bunsen flame. The end in the hand will soon begin to grow warm. An examination of the rod will show that all its parts are heated, the more

---

[1] Always cut phosphorus under water, and never handle it with dry fingers.

highly as the flame is approached. Hold the hand above the flame; it will be warmed by a current of warm air rising from the flame. Hold the hand by the side of the flame; again a sensation of warmth will be perceived.

In this experiment we recognize three distinct modes by which energy in the form of heat is transmitted from one point to another:

1. *Conduction*, in which heat is communicated from each molecule to the adjacent ones, as in the case of the metallic rod.

2. *Convection*, in which heat is transferred by transferring the heated particles of some kind of matter, as in the case of the heated air rising and warming the hand.

3. *Radiation*, in which one body loses and another gains heat by means of a process of such a nature that the intervening medium is not heated; such was the case when the hand was warmed at the side of the flame, the intervening air not being perceptibly affected.

196. Conduction. — Exp. — Select two stout wires, iron and copper, of the same gauge-number, each about 40 cm. long. Twist them together for about 10 cm., and then separate the untwisted parts, bending them to give the apparatus the form of a fork. Support the wires on a wooden frame, and apply heat to the twisted part.

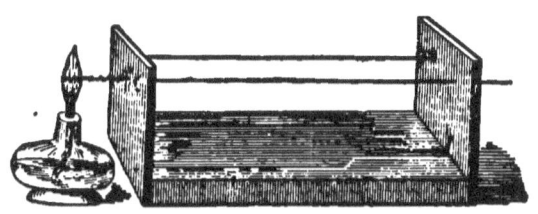

Fig. 90.

After several minutes find the point on each rod farthest from the flame where an ordinary match ignites on touching it. This point will be found to be some distance farther along on the copper than on the iron, apparently showing that copper is a much better conductor of heat than iron.

Exp. — Fasten a cylinder of wood into a brass tube, so that the two have the same outside diameter. Wrap a piece of writing-paper tightly around the junction and hold it in a flame (Fig. 91). The paper in contact with the wood soon burns, whereas that in contact with the brass is not injured. The metal conducts away the heat so rapidly that the paper around it does not reach the temperature of ignition as soon as that around the poorer conductor.

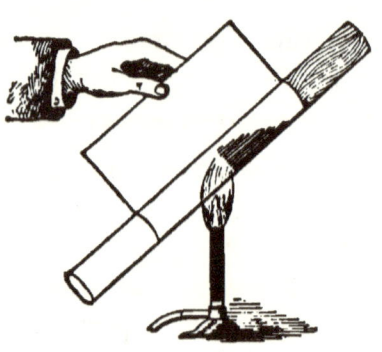

Fig. 91.

Hence, we see that substances differ in ability to conduct heat.

Conclusions as to *relative* conductivity, however, are likely to be erroneous, if, in our experiments, we assume that it takes the same amount of heat to raise equal quantities of two substances through the same difference of temperatures, and that radiation goes on alike from both. The poorer conductor may appear to be the better. Such would be the case with lead and iron, if we should compare them by fastening balls to bars of each by means of shoemaker's wax, and then record the time necessary to release the balls when heat is applied to the ends of these bars. The heat that will raise a mass of lead four degrees will raise an equal mass of iron only about one degree. Hence, the lead will acquire the temperature necessary to melt wax and release the balls long before the iron; and although it is in reality the poorer conductor, the experiment would give it the appearance of being the better conductor.

Among solids the metals are the best conductors; substances in a powdered state are poor conductors, owing to a lack of continuity in the material. Wood, leather, flannel, and organic substances generally are poor conductors.

197. Conductivity of Liquids. — **Exp.** — Pass the stem of an air thermometer through a cork fitted into the neck of a large funnel. Support the apparatus as shown in Fig. 92, adjusting it so that the lower end of the stem dips into a vessel of colored water. Fill the funnel with water, covering the bulb to the depth of about two centimetres. Support on the top in a wire frame a small porcelain dish with its bottom dipping into the water directly over the bulb. Pour a spoonful of alcohol into the dish and set it on fire. By observing the position of the colored water in the tube, it will be seen that the intense heat produced by the burning alcohol has little or no effect on the bulb of the thermometer. On stirring the water the liquid index moves rapidly downward.

Fig. 92.

It thus appears that water is a very poor conductor of heat. It is very doubtful if water, and, in fact, most liquids, conduct heat at all, the distribution of heat through them being due to currents set up in them (200) and also to diffusion.

198. Conductivity of Gases. — On account of radiation it is almost impossible to determine whether gases have any conducting power. The probability is that they are absolute non-conductors, and that any heat which they transmit is by currents produced in them and by diffusion.

199. Applications. — Some articles in a room feel cold to the touch when others feel warm. An examination of the matter will reveal the fact that those which feel cold are good conductors of heat, whereas the warm ones are bad conductors. The former lead the heat away from the hand faster than the body supplies it, causing the sensation of cold; the latter do not carry off the heat, and consequently there is a sensation of warmth.

The handles on metallic instruments are usually made of some poor conductor, as wood, bone, etc., or else are insulated by the insertion of some non-conductor, as in the case of the handles to silver tea-pots, where pieces of ivory are inserted to keep them from becoming heated. A small coil of copper wire held in the flame of a candle will frequently extinguish it by conducting away the heat, thereby lowering the temperature below the point of combustion.

The non-conducting character of air is utilized in houses with hollow walls, in double doors and double windows, and in clothing of loose texture. The warmth of woollen articles and of fur is due mainly to the fact that much air is enclosed within them on account of their loose structure.

An interesting illustration of the high conductivity of copper is seen in the fact that a copper tea-kettle begins to "sing" at a lower temperature than an iron one. The heat passes through the copper more readily, and causes bubbles of vapor to form more rapidly. These on rising through the cold liquid above collapse in such rapid succession that a musical sound is produced. If we boil water in a glass vessel no singing is noticeable, owing to the slowness with which heat passes through the glass, and to the smoothness of the internal surface.

HEAT.

EXERCISES.

1. Why do not snow and ice on melting liquefy at once?
2. Why does a metal liquefy so rapidly on beginning to melt?
3. Why does snow protect from cold?
4. Why will a current of air extinguish flame?
5. Account for the non-melting of ice when packed in sawdust?
6. Why do men working about smelting-furnaces wear flannel clothing?
7. Account for the slowness with which ice increases in thickness over a pond.
8. What explanation would the molecular theory of matter suggest for the poor thermal conductivity of fluids?
9. Why are many culinary vessels made with copper bottoms?
10. Why is it difficult to boil water in a " furred " kettle?
11. Why is paper so effective in protecting plants from frost?

200. Convection. — It has already been pointed out that a liquid expands on being heated, and hence has its density lessened. Why? Therefore, it follows that if any portion of a liquid in the lower part of a vessel be heated, it will rise in obedience to the principle of flotation (167). The same is true in the case of gases. The currents produced in this way are called *Convection Currents*.

Exp. — Construct out of glass tubing a rectangle about 50 cm. long by 15 cm. wide. At one corner, by means of rubber connectors, insert a T-tube, for convenience in filling, and also as a provision for expansion (Fig. 93). Fill the apparatus with water, freed from air by boiling, and introduce a few paper raspings. Now hold one corner of the rectangle in a vessel of boiling water. The particles of paper will soon begin to move around the rectangle. By applying ice to the side of the rectangle in which the current is ascending, the direction of the flow may be reversed. Explain.

134   *ELEMENTS OF PHYSICS.*

Buildings and railway coaches are frequently heated by the circulation of hot water through pipes. Many of the ocean currents are illustrations of convection.

Exp.—Fill a large glass jar or beaker about three-fourths full of cold water, and pour on it carefully some warm water colored with

Fig. 93.

aniline blue, in sufficient quantity to form a layer about 2 cm. in thickness. Hold in the colored water a lump of ice; this will cool the surface layer, and presently streams of colored water will be seen descending through the uncolored liquid. Explain.

Exp.—Heat to redness a piece of sheet iron. Hold above it a set of paper vanes balanced on a bent wire in such a way as to rotate freely. The heated air rises, strikes against these vanes and causes them to rotate rapidly. Why? To make such a set of vanes, cut a disk of cardboard about 10 cm. diameter. Make, say, eight radial cuts in it, not reaching quite to the centre. By bending the material, set each sector obliquely to the plane of the circle.

By burning touch-paper[1] it will be seen by the movement of the smoke that there is a movement of the air on all sides toward the space directly over the heated plate.

201. Ventilation. — **Exp.** — Support in a shallow dish a short piece of candle and place over it a lamp-chimney. Pour enough water into the dish to close the lower end of the chimney. The flame is soon extinguished. Why? Relight the candle, and insert a cardboard partition in the chimney, as in Fig. 94. The candle will now continue to burn, and if we hold a piece of lighted touch-paper over the top of the chimney we shall find that there is a current of air down one side and up the other.

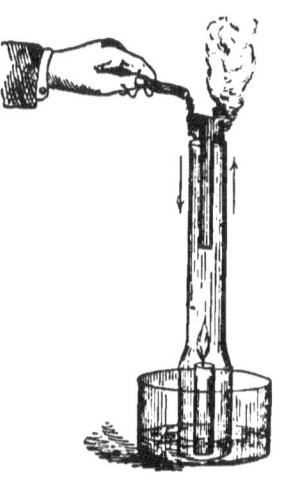

Fig. 94.

The office of a chimney is to increase the supply of oxygen to the flame. The air within it being heated by the flame rises, and cold air moves in through the bottom to restore the equilibrium, becomes heated in passing over the fire, and thereby maintains the high temperature. (Why does increasing the height of the chimney increase the draught?) This principle underlies most methods of ventilation of rooms. A flue in the wall carries off the impure air of the room, and air flows into the

---

[1] Made by soaking porous paper in a strong solution of saltpetre and drying. It burns, giving off considerable smoke.

room to take its place, passing over heated pipes or a heated furnace in so doing.

Sometimes flues fail to work because they are too wide; in which case the cold air may flow down one side and up the other just as if there was a partition in it. They may also fail to work because they do not open at the highest point of the building, in which case the pressure will be reversed and the current in the flue will be downward. Why? More frequently the failure is caused by the flue being placed in an outside wall, and hence the air within it is no warmer than that without; in which case the air will flow down the flue into the room, because the air of the room is less dense than that without. Why?

That the existence of flues opening into a room does not ensure ventilation, unless means are adopted to make certain the upward movement of the air in such flues, is illustrated in the following experiment:

Exp.—Fit to a wide-mouthed bottle of about 2 l. capacity a cork through which pass two glass tubes, each at least 2 cm. diameter and 20 cm. in length, the upper and the lower ends of these tubes being at the same level. A wire also passes through the cork, carrying a piece of candle at the lower end. The shape of the wire is such that the candle can be brought directly under either tube, or can be turned away from both of them. First, set the candle in the latter position, light it, and insert the cork with its tube in the bottle. The flame will soon go out, no air entering through either tube, although both are open. Secondly, blow out the foul gas, relight the candle, turn the wire till the flame is directly under one of the tubes, and insert the cork in the bottle. The candle will continue to burn brightly. If we apply lighted touch-paper to the top of these tubes in succession, we shall

Fig. 95.

find that there is a downward current in one and an upward one in the other.

202. Radiation. — A person standing near a hot stove experiences a sensation of warmth. The heat received is not communicated by contact, neither is it conducted through the air. Why? Physicists are of the opinion that the rapidly vibrating molecules of the heated body set up in a subtle medium called *the ether*, which is believed to fill all space, a variety of wave motion. When these waves strike our bodies they produce the sensation of heat. Heat propagated in this way is, for convenience, referred to as *Radiant Heat*, although it does not pass from point to point as heat, but rather as radiant energy transformable into heat by absorption. It is in this way that heat reaches the earth from the sun.

203. Laws of Heat Radiation. — Experiments prove the following laws: I. *Radiation proceeds in straight lines.* This is illustrated in the use of fire-screens and sun-shades.

II. *The amount of radiant energy received by a body varies inversely as the square of the distance from the source.*

III. *Radiant energy is reflected from a polished surface so that the angles of incidence and reflection are equal.* Archimedes is said to have set fire to the Roman ships during the siege of Syracuse in 212 B.C., by concentrating on them the heat of the sun by the aid of a large concave reflector.

IV. *The power of substances to reflect radiant energy depends both on the polish of the surface and the nature of the material.* Polished brass is the best reflector, and lampblack is the poorest.

V. *The rate at which the temperature of a cooling body falls by radiation is proportional to the excess of its temperature over that of the surrounding medium.* This is known as *Newton's Law of Cooling*, and is very nearly correct where the excess is under 20° C.

204. Absorption of Heat. — **Exp.** — In slots, 10 cm. apart, cut in a narrow board, support two pieces of bright tin plate, each 10 cm. square. Coat the inner face of one of these squares of tin with lampblack. Stick balls of equal size at the centre of each outside face with shoemaker's wax, using as little as possible. Hold a heated block of iron midway between the two plates. The ball will soon fall from the blackened plate, showing that lampblack is a ready absorbent of heat.

By coating one of the plates successively with different substances, it will be found that *substances differ in their power of absorbing heat.* Leslie discovered that the best absorbers, as lampblack, ashes, rough surfaces, are bad reflectors; while good reflectors, as polished metals, are bad absorbers.

205. Selective Absorption. — **Exp.** — Fill a large flat bottle, about 3 cm. thick, with clear water; place it between the sun and the bulb of an air thermometer, and compare with the effect obtained when nothing intervenes. Repeat the experiment, using in place of the water a solution of iodine in carbon disulphide.

A comparison of the results reached will show that clear water cuts off nearly all radiant heat, whereas the solution of iodine does not perceptibly affect the intensity. Substances which transmit radiant heat are called *Diathermanous*, and those which do not, *Athermanous*. Rock-salt is the most highly diathermanous substance known. On the other hand, alum, sugar, glass, water, and ice are extremely athermanous. The diathermanous character of a

HEAT.

substance varies with the nature of the radiant. Such substances as alum, water, etc., transmit little or none of the radiation from a surface of low temperature, but allow considerable to pass when the radiating body is one of high temperature. The radiant energy from the sun passes readily through the atmosphere to the earth, warming its surface; but that from the earth is stopped by the enveloping atmosphere. Likewise, the radiant heat from the sun passes readily through the glass of the greenhouse, but that from within is unable to pass outward.

EXERCISES.

1. The moon has no atmosphere. What effect must this have on the monthly range of temperature on its surface?
2. If Newton's law of cooling be true, calculate the following: The temperature of a room is 60° F., and a thermometer cooling in it indicates the readings 180° and 140° at an interval of one minute. What will be its reading after another minute?
3. On the top of a high mountain a person is blistered in the sun and frozen in the shade. Explain.
4. Is brick or polished tile the best substance for the back of a fire-place?
5. Which will heat the sooner in the sun, a polished or a rough piece of brass? Explain.
6. Large tin cans are employed in carrying milk. It is desirable to keep the milk as cool as possible. Would it be wise to paint the cans on the outside?
7. Ought the bottom of a tea-kettle to be polished?
8. Account for winds.
9. Why does snow near a tree melt sooner than in the open field?
10. How are safes made fire-proof?

V. EFFECTS OF HEAT.

206. Expansion of Solids. — **Exp.** — Fasten a metal rod about 30 cm. long to a heavy wooden block. Let the other end of the rod rest against a pointer pivoted to a vertical board. The bearing should be as near the pivot as possible. Place a flame under the rod. The moving pointer will indicate that the rod is increasing in length.

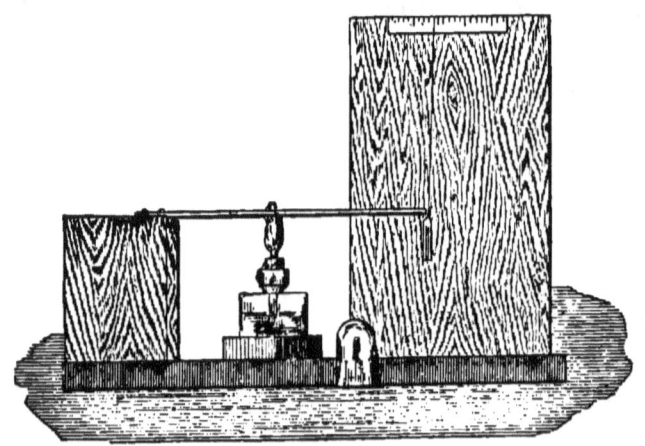

Fig. 96.

Exp. — Cut a round hole in a sheet of tin so that a selected metallic ball will just pass through it. On heating the ball it will be found that it is too large for the opening, showing that it has expanded in every direction.

Exp. — Rivet together at intervals of 2.5 cm. a strip of sheet iron and one of copper, each about 20 cm. long. Heat the middle of this compound bar in a flame. In a few minutes it will be appreciably curved, with the copper on the convex side, because the two metals expand unequally.

Such experiments as these establish the following laws:

I. *Solids expand when heated and contract when cooled.* Iodide of silver is the only known exception.

II. *The amount of expansion varies with the substance.*

HEAT. 141

207. Expansion of Liquids. — Exp. — Select several glass tubes of the same bore, and each about 15 cm. long. Close one end of each by fusion. Fill each to the same height with some liquid, as water, alcohol, glycerine, etc. Hold them in a vessel of hot water. The liquids will rise in the tubes, but not to the same level.

Hence, we conclude that *liquids expand, the amount varying with the substance.*

208. Expansion of Gases. — Exp. — Select two Florence flasks of the same shape and volume. Fit to each, air tight, a perforated stopper, through which is a bent delivery-tube. Fill one

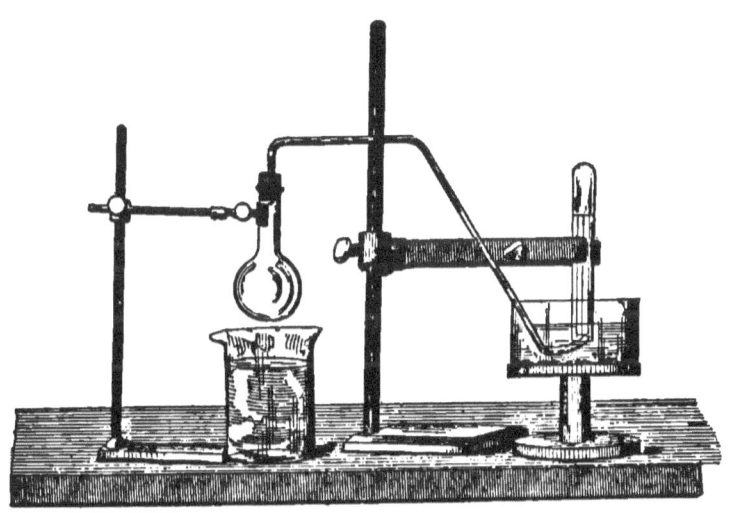

Fig. 97.

flask with air and the other with coal-gas. Insert each flask to the same depth in a vessel of hot water. The gases, expanded by the heat of the water, escape through the delivery-tubes. Collect the escaping gases in graduated test-tubes over water. On removing the water-bath, the gases cool and water enters the flasks. A careful inspection of the test-tubes will show that they contain about equal quantities of gas. Hence:

I. *Heat expands gases and cold contracts them.*

II. *For a given change in temperature, gases suffer a greater change of volume than either liquids or solids.*

III. *All gases expand at the same rate.*

209. Applications. — Many familiar phenomena are explained by expansion. If hot water be poured into a thick glass vessel, it will break because of the strain due to the sudden expansion of the inner surface. Glass is a poor conductor of heat, and hence glass vessels which are to be subjected to extremes of temperature must be thin, in order that the two surfaces may not differ much in temperature and cause unequal expansion. In laying the iron rails of a railroad in cold weather, spaces are left between the ends to permit of expansion. We have already seen that in the thermometer the principle of expansion is utilized, or rather the principle of differential expansion, for if the mercury and the glass expanded alike, the instrument would not work. Why? The metallic thermometer utilizes the unequal expansion of metals; a compound bar, as it changes its shape with the changes of temperature, moves a pointer over a graduated scale. Glass and platinum expand very nearly at the same rate. For that reason platinum wires can be fused into glass vessels and not crack out on cooling.

210. Coefficient of Expansion. — It has been shown that substances when heated expand in every direction. This is called *Cubical Expansion*, in distinction from *Linear Expansion*, or expansion in one direction only. *The Coefficient of Linear Expansion* of a body is the fraction of its length which a body expands when heated from 0° to 1° C. *The Coefficient of Cubical Expansion* is the fraction of its volume which a body expands when heated from 0° to 1° C.

HEAT. 143

210 a. The Law of Charles. — The volume of any mass of a gas, *under constant pressure*, increases from the freezing to the boiling point by a constant fraction of its *volume at zero degrees*. This law is known as the law of Charles. For the centigrade scale the constant fraction is 0.3665 for dry air. For 1° C. this fraction or coefficient of cubical expansion is therefore 0.003665. A near approximation is $\frac{1}{273}$. Hence, 30 c.c. at 0° C. become about 41 c.c. at 100° C. If v_0 is the volume at 0° C., then at any other temperature t the volume becomes $v = v_0(1 + kt)$, where k is the coefficient of cubical expansion. This law, like that of Boyle (152), is not rigorously exact; but the approximation is closer and closer as gases are more highly rarefied.

The law of Charles leads to a fourth scale of temperature, called the *absolute scale*. If it be assumed that the law holds for very low temperatures, that is, if the volume continues to diminish $\frac{1}{273}$ for every degree fall of temperature, then at $-273°$ C. the volume of the gas would be reduced to zero. This must be regarded as the limit toward which the volume tends. At this low temperature the molecules have lost all their motion, and the gas has lost all its heat energy. To convert centigrade readings into those of the absolute scale, add 273.

It follows from the law of Charles that the volume of a gas, under constant pressure, is simply proportional to its temperature measured on the absolute scale. Thus, 100 c.c. of air at 27° C. becomes 121 c.c. at 90° C.; for these two temperatures on the absolute scale are 300° and 363° respectively, and 100 and 121 have the same ratio as 300 and 363.

211. Expansion on Crystallizing. — **Exp.** — Fit to a small bottle a perforated cork through which passes a glass tube. Fill the apparatus half way up the inserted tube with water, freed from air by boiling, and then pack in a mixture of finely broken ice and salt. The water-column will slowly fall for a few minutes, and then it will begin to rise till water flows out of the top of the tube. On examination it will be found that the water is frozen.

Those substances which crystallize on cooling as a rule expand when their temperature approaches that of the solidifying-point. Crystalline structures occupy more space than the same matter in a liquid form.

212. Force of Expansion and Contraction.—Exp.— Pour about one ccm. of water into a small test-tube and close the end by fusion. Lay it in an empty sand-bath on the ring of the iron stand. Apply heat and stand at a safe distance. In a few minutes there will be a loud report due to the bursting of the tube.

The force of expansion or of contraction of a substance is evidently equal to that necessary to compress or expand it to the same extent by mechanical means, and hence can be computed by proceeding in the manner illustrated in the following example: A bar of malleable iron, one square inch in cross-sectional area, if placed under a tension of one ton, increases in length .0001 of itself. The coefficient of linear expansion of iron is .000012204. Since $.0001 \div .000012204 = 8 +$, a change of temperature of about 8° C. will produce the same change in the length of the bar as a force of one ton.

Applications of this great contractile force are numerous. In fitting tires to wheels, they are cut somewhat smaller in diameter than the wheels, then expanded by heat till they slip on. On cooling they contract and bind the parts firmly together. In riveting together the plates of a steam-boiler, the bolts are heated so that on cooling the plates may be forced together and the joints made water-tight. In all heavy metal structures, such as railroad bridges, one end at least must have freedom of motion; otherwise, the changes of length during variations of temperature would result in the deformation of the structure.

EXERCISES.

1. Why will heating the neck of a bottle frequently loosen the glass stopper?
2. Why does a clock lose time in summer and gain in winter?
3. A rod of zinc measures 1 ft. 3 in. at $0°$ C.; what will it measure at $6°$ C.? (Coefficient of expansion of zinc $= 0.00002976$.)
4. A cube of copper measures 1 ft. on each edge at $15°$ C. Find its volume at $180°$ C. (Coefficient of expansion of copper $= 0.00001866$.)
5. Why does temperature affect density?
6. Water has its greatest density at $4°$ C. If a pond is freezing over, what is the temperature of the water at the bottom?
7. Find the coefficient of expansion of a rod of brass from the following data: at $15°$ C. it measures 2 ft. long; at $95°$ C. it measures 2.003 ft.
8. At $0°$ C. the volume of a certain mass of air is 1092 cu. ft. What will be the volume at $20°$ C., and at $-20°$ C.?
9. The coefficient of expansion of brass is .000019. A certain distance measured with a brass yardstick is 420 ft. 6 in. The temperature was $21°$ C.; the yardstick was correct at $16°$ C. Find the correct distance.
10. 15 litres of oxygen at $10°$ C. will have what volume at $0°$ C.?
11. Three gallons of gas at $10°$ C. shrink to $2\frac{1}{2}$ gallons at what temperature?
12. A litre of air at $0°$ C. and 760 mm. pressure weighs 1.293 gram.; what will a litre of air at $15°$ C. and 763 mm. weigh?

213. **Liquefaction.** — One of the most familiar effects of heat is the change in the state of the substance. When a body passes from a solid to a liquid condition it is said to *Melt* or *Liquefy*, and the act of changing is called *Lique-*

faction. The temperature at which this change occurs is the *Melting-Point.*

214. Solidification. — Nearly all liquids may be changed to the solid form by lowering their temperatures, that is, they *Freeze* or *Solidify.* The temperature at which this change occurs is usually the same as the melting-point. Some substances, as water and sulphur, if undisturbed, can be cooled several degrees below their true solidifying-point and still remain liquid. On the slightest disturbance, however, solidification at once sets in and the temperature rises to that of the usual solidifying-point.

215. Laws of Fusion. — It is found experimentally that the fusion of solids is in accordance with the following laws:

I. *Every crystalline substance begins to melt at a certain temperature, invariable for each substance, if the pressure be constant.*

II. *The temperature of a substance when slowly melting remains constant till all is melted.*

III. *Substances which contract on melting have their melting-points lowered by pressure, and vice versa.*

216. Disappearance of Heat during Liquefaction. — During the process of liquefaction there is no rise of temperature, although heat is constantly applied to the substance. When this fact was first observed, it was generally believed that heat was a kind of matter. This led to the introduction of two terms, *Sensible Heat* and *Latent Heat,* the former denoting that form of heat which affects the temperature; and the latter, that form which it was supposed to assume when it no longer affects it. These terms have been retained, notwithstanding the falsity of the materialistic theory of heat, and definitions in harmony with the kinetic theory have been given them.

In order that a substance may pass from the solid to the liquid state, work must be done against the force of cohesion. To effect this, energy must be expended, and this is supplied by the heat which is communicated to the substance. Hence, the energy transferred to the body is in part transformed into the potential form, which does not increase the kinetic energy of the molecules, and does not cause an increase of temperature. Therefore, it appears that when heat is applied to a body, part goes to increase the molecular *kinetic energy*, which appears as *Sensible Heat;* and part goes to increase the molecular *potential energy*, that is, becomes *Latent Heat* (235).

217. Disappearance of Heat during Solution.— Exp. — Pour a few cubic centimetres of water into a beaker and ascertain its temperature. Then add a few crystals of sodium sulphate. The temperature will fall as they dissolve.

Just as in melting, the cohesion of the substance has to be overcome in effecting the solution, and hence some of the kinetic energy is transformed into potential energy, or, in other words, heat disappears. It will often happen that this absorption of heat is disguised by the heat evolved by chemical action between the substances.

218. Evolution of Heat during Solidification. — When a liquid solidifies it must part with the energy necessary to maintain its liquid condition against cohesion. This it does in the form of sensible heat. The potential heat of liquefaction then changes to the kinetic form when solidification takes place, the amount of heat now appearing being evidently equal to that which disappears during fusion.

An application of the above principle is seen in protecting cellars from frost by placing in them tubs of water. The water on freezing gives off heat, which maintains the temperature sufficiently high to protect the vegetables. The amelioration of the cold of winter and the heat of summer by large bodies of water is similarly accounted for.

219. Freezing Mixtures. — **Exp.** — Mix together one part by weight of common salt, and two parts of snow or pounded ice. A thermometer placed in the mixture will probably indicate a temperature as low as — 20° C.

Exp. — Powder together equal weights of ammonium chloride and potassium nitrate, and dissolve in twice their weight of water. A thermometer in the solution will show a fall of about 20° in temperature.

All such freezing mixtures are based on the principle that heat is absorbed (rendered latent) in the passage of bodies from the solid to the liquid condition.

220. Vaporization is the act of changing a substance into a gaseous condition. If this change takes place slowly, without producing any noticeable disturbance in the body of the substance, it is called *Evaporation*. If, during the act, the body is visibly agitated by rapid internal evaporation, it is called *Ebullition*.

221. Illustrations. — **Exp.** — Pour a few drops of ether into a beaker and cover it loosely with a plate of glass. Let it stand for a few moments, then bring a lighted match to the mouth of the vessel. A sudden flash will show that the vapor of ether filled the vessel.

Exp. — Support on the iron stand a beaker two-thirds full of water. Apply heat; in time bubbles of steam will rise from the bottom of

HEAT. 149

the beaker, through the liquid, and burst at the top. The water is now "*boiling.*"

222. Laws of Evaporation. — Experiment shows that evaporation takes place in accordance with the following laws:

I. *The rate of evaporation increases with increase of temperature.*

II. *The rate of evaporation increases with the free surface of the liquid.*

III. *The rate of evaporation is accelerated by a continual change of the air in contact with the liquid.*

IV. *The rate of evaporation is accelerated by diminution of pressure.*

223. Explanations and Applications. — That a large surface should be favorable to evaporation is easily accounted for by the fact that a great number of points is afforded for the formation of vapor. The principle is utilized in the manufacture of salt by having large shallow pans for the brine, or by having the brine trickle over large bundles of twigs.

The effect on the rate of evaporation due to changing the air next to a liquid has a most familiar illustration in the action of wind in drying the roads after a rain, and in drying wet articles when hung up. Were the surrounding air at rest, it would soon become saturated with moisture, and the molecules in their erratic movements would return in as great numbers to the body from which they came as those escaping from it, thus neutralizing any loss from evaporation. This is prevented when the air is in motion.

In order that syrups may be concentrated at a low temperature to avoid burning, the operation is carried on in large covered pans from which the air and the vapor are exhausted by air-pumps. The space above the liquid does not become so crowded with molecules as to impede the movements of the molecules leaving the surface of the liquid.

224. Laws of Ebullition. — Experiment reveals the following laws regarding the boiling-point of a liquid:

I. *Each liquid has its own boiling-point, which is invariable for that substance under the same conditions.*
II. *The boiling-point is dependent upon the character of the surface of the containing vessel.*
III. *The boiling-point is raised by salts and lowered by gases dissolved in the liquid.*
IV. *The boiling-point increases with the pressure.*

225. Franklin's Experiment. — Fit to a Florence flask a delivery-tube as shown in Fig. 98. The bore of the tube should be small, and the cork should fit airtight. Fill the flask about one-third full of water and bring it to the boiling-point. After the boiling has continued for a few minutes, remove the lamp, and let the delivery-tube dip into a small dish of water. As the water in the flask cools, that in the beaker will rise in the tube. Why? In a few minutes it will begin to flow over into the flask, and a violent boiling of the water in the flask will take place.

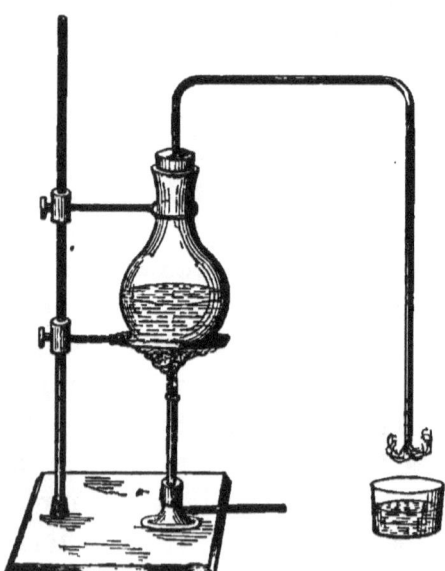

Fig. 98.

The introduction of the cold water condenses the vapor above the water in the flask and lowers the pressure, with the result that the water boils vigorously, though its temperature is considerably below that of the true boiling-point.

226. Measurement of Heights by the Boiling-Point. — Since atmospheric pressure decreases with the elevation, it is evident that the boiling-point of a liquid must also decrease. Hence, the boiling-point of water may be used as an indicator of the height of a place above the level of the sea. A close approximation to the altitude of a place is obtained by allowing 295 metres for each degree C., or 538 ft. for each degree F., that the observed boiling-point of water is below the boiling-point at sea level. Thus, at Quito, the highest city in the world, where the average boiling-point of water is 90.1° C., the height above sea level would be $295 \times 9.9 = 2920.5$ metres, a quantity greater than the true height by 34.4 metres.

227. Disappearance of Heat during Evaporation. — If a thermometer be suspended in the vapor above a boiling liquid, its reading is constant so long as the pressure remains unchanged, although large quantities of heat are continually imparted to the substance. The large amount of heat rendered latent is accounted for by the work done in separating the molecules beyond the range of cohesion, and overcoming the pressure of the air and vapor above the liquid (236).

228. Cold by Evaporation. — **Exp.** — Pour a few drops of ether on the bulb of an air thermometer. A rapid fall of temperature will result.

In the vaporization of the ether work is done, some of

the kinetic energy of the air in the bulb is transformed into potential energy, that is, part of its heat is used to vaporize the ether, thereby lowering its temperature. Many applications are made of the principle that evaporation is attended by a lowering of temperature. The rapid evaporation of liquid ammonia is utilized in the artificial production of ice. Sprinkling the floor of a room cools the air, because of the heat rendered latent by the evaporation of the water. Professor Dewar has recently liquefied oxygen on a large scale by means of a very low temperature, obtained by the successive evaporation of liquid nitrous oxide and ethylene.

229. Heat by Condensation. — When a vapor is liquefied all the heat that disappeared during vaporization is again generated or rendered sensible. Application is made of this fact in steam-heating.

230. Distillation is an application of the principles of evaporation and condensation. Fractional distillation consists in the separation of two or more liquids of different boiling-points from one another. The apparatus used for vaporizing the liquids is called the *Still;* and that for liquefying,

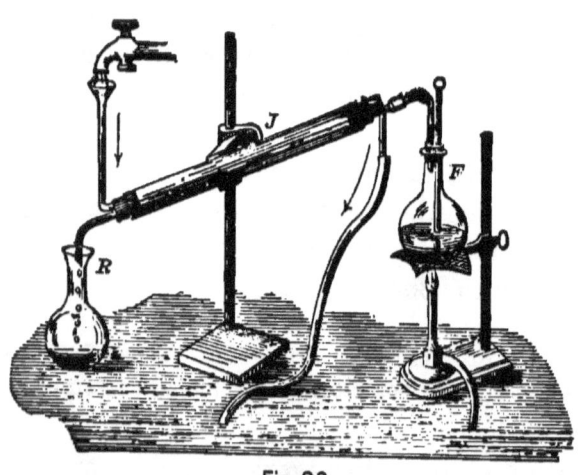

Fig. 99.

the *Condenser*. In practice, the condenser frequently consists of a coiled tube, called the *Worm*, surrounded by water. The following experiment illustrates the process as carried on in laboratories:

A mixture of alcohol and water is placed in the flask F (Fig. 99). The delivery-tube passes through a large tube J, which is supplied with cold water, as shown in the figure. By keeping the flask at a temperature intermediate between the boiling-points of water and alcohol, the vapor which escapes will be largely that of alcohol. This will be condensed by the cold water surrounding the delivery-tube and collected in the flask R.

EXERCISES.

1. How can pure water be obtained from sea water?
2. How can water be heated above the ordinary boiling-point?
3. How high is a mountain where water boils at 85° C.?
4. Mt. Washington is 6,288 feet above the sea-level; at what temperature will water boil on its top?
5. Hang a thermometer under the receiver of an air-pump, exhaust the air, and account for the falling temperature.
6. Why will a thermometer with moist bulb register lower in the wind than when protected?
7. Pour water at the temperature of the room into an unglazed earthen vessel. Insert a thermometer. Account for the lowering of the temperature.
8. Account for the low temperature on the tops of mountains.
9. Why is an iceberg frequently enveloped in a fog?

VI. CALORIMETRY.

231. Terms used. — The quantity of heat necessary to raise the temperature of unit mass of water 1° C. is the *Thermal Unit*. If a gramme is the unit of mass, the thermal unit is called a *Calorie*. Any unit of mass may be taken as a standard in fixing the magnitude of the thermal unit, provided this unit of mass is not changed throughout any problem. *The Thermal Capacity* of a body is the number of thermal units necessary to raise its temperature 1° C. *The Specific Heat* of a substance is the thermal capacity of unit mass of the substance. For example, the specific heat of mercury is 0.0335, meaning that the heat which will raise the temperature of unit mass of water 0°.0335 C., or 0.0335 of unit mass 1° C., will raise the temperature of unit mass of mercury 1° C. (How many times greater is the thermal capacity of a gramme of water than that of a gramme of mercury?)

232. Thermal Capacity of Water. — Exp. — Select two thin glass beakers of about one litre capacity each. Pour into each 400 grammes of water, one at the temperature of the room and the other at about 60° C. Now pour the hotter of the two into the colder. The temperature of the mixture will be about the arithmetical mean of the temperatures of the two vessels before mixing. For example, if the temperatures were 20° and 60° respectively before mixing, that of the mixture will be about 40°, the small discrepancy being due to radiation and absorption by the vessel.

Hence, we see that the heat which raises a quantity of water from 40° to 60° is sufficient to raise an equal quantity from 20° to 40°; or the quantity of heat which will raise a quantity of water 20° at one part of the thermometric scale will raise an equal amount 20° at another part of the scale, a result which would not be reached, *if the specific heat of water were not approximately the same at all temperatures.*

233. Determination of Specific Heat. — Exp. — Make a loose coil of sheet lead, weigh it, and suspend it by a thread for several minutes in boiling water. Take the temperature of the water and you will have that of the lead. Weigh out in a beaker a quantity of water sufficient to cover the lead and bring it to the temperature of the room. Now transfer to it the coil of lead from

HEAT. 155

the boiling water, stir the water gently with a thermometer and record the temperature as soon as it ceases to rise. The number of thermal units gained by the water (which is the number lost by the lead) will be the product of the mass of the water by the gain in temperature. This divided by the lead's loss of temperature will be the thermal capacity of the lead, and this by the mass of the lead will be its *specific heat*. Why?

The above method of finding the specific heat of a substance is known as the *Method of Mixture*. To obtain accurate results the data obtained must be corrected for radiation and absorption by the vessel.

234. One Cause why Substances differ in Specific Heat. — When heat is applied to a body, part goes toward raising its temperature, and part is consumed in doing *internal* and *external* work, or in doing work against cohesion and outside pressure in giving new positions to the molecules. Since in different substances the force of cohesion differs considerably, we should expect to find that varying amounts of heat-energy are spent in doing work against it. This becomes latent, and consequently not the same quantity remains in each substance to produce changes in temperature.

235. The Latent Heat of Fusion is the number of thermal units of heat required to melt unit mass of the substance without raising its temperature. The following experiment shows how it may be obtained for water:

Pour into a beaker 500 grammes of water at a temperature of 60° C. Add to it, in small pieces, 200 grammes of ice. Take the temperature of the water as soon as the ice is melted. Suppose it is 20°. The number of calories of heat consumed will be 500 times the fall of temperature, that is, $500 \times 40 = 20000$. But 200 grammes of ice-water were raised from 0° to 20°. Hence, $200 \times 20 = 4000$ calories were expended in heating the ice-water up to 20°; while

20000 — 4000 = 16000 calories were required to melt 200 g. of ice. Therefore, 16000 ÷ 200 = 80 calories is the amount of heat required to melt one gramme of ice.

When corrections for radiation and absorption are made it is found that the heat of fluidity of water is 79.25.

236. The Latent Heat of Vaporization is the number of thermal units required to change unit mass of the substance at its boiling-point into vapor at the same temperature. It may be found for water as follows:

Set up an apparatus like that shown in Fig. 100. Boil water in the flask and convey the steam into a beaker containing a known quantity of water. The increase in the weight of the water gives the amount of steam condensed, and the increase of temperature gives the amount of heat given off. The "trap" in the delivery-tube catches the water that condenses before reaching the beaker. Suppose that the experiment furnishes the following data: Amount of water in the beaker 400 g. at the beginning, 414 g. at the end. Temperature at the beginning 20° C., at the end 40° C. Observed boiling-point 99° C. Then, there were 14 g. of steam introduced, which first condensed to water at 99° C., and afterwards fell 99° — 40° = 59°. 400 g. of water increased 20°, for which 400 × 20 = 8000 calories were required. 14 × 59 = 826 calories came from the condensed water in cooling from 99° to 40°. Hence, 8000 — 826 = 7174 calories came from the steam in condensing to water at 99°; 7174 ÷ 14 = 512.4, the number of calories given out by one gramme of steam in condensing.

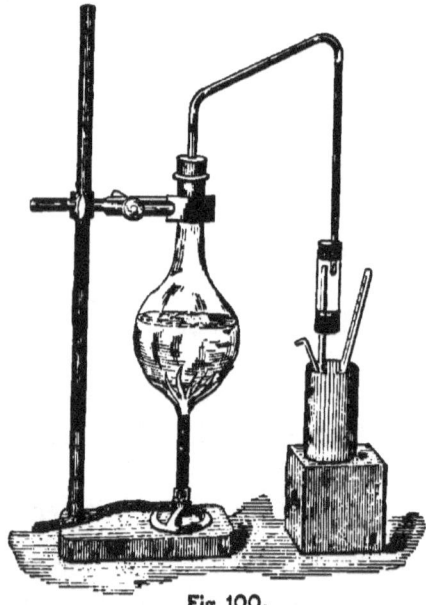

Fig. 100.

When corrections for absorption and radiation are made it is found that the latent heat of steam is 535.9.

EXERCISES.

1. Equal masses of boiling water and of mercury at $-5°$ C. are mixed together with a resulting temperature of $96°.65$ C. Find the specific heat of mercury.
2. How much ice at $0°$ C. will be melted by 500 grammes of boiling water?
3. How much ice must be dissolved in a litre of water at $20°$ C. in order to reduce its temperature to $5°$ C.?
4. Find the result of mixing 2 lbs. of ice at $0°$ C. with 3 lbs. of water at $45°$ C.
5. 30 g. of iron nails at $100°$ C. are dropped into 60 g. of water at $13°.2$ C., and the final temperature is $18°.6$ C. What is the specific heat of the nails?
6. How much heat is required to raise 150 g. of copper (sp. ht.$=$ 0.095) from $10°$ to $150°$ C.?
7. 120 g. of ice melted in 300 g. of water at $50°$ C. reduced the temperature of the water to $13°$ C. Compute the latent heat of fusion of water.
8. 10 g. of steam at $100°$ C. condensed in one kilogramme of water at $0°$ C. raised the temperature to $6°.3$ C. Calculate the latent heat of steam.
9. How much steam at $100°$ C. is required to raise 150 g. of water from $0°$ to $100°$ C.?
10. How much heat would it require to raise 250 g. of ice from $0°$ C. to $100°$ and to convert it into steam?

VII. HEAT AND WORK.

237. Relation between Heat and Work. — By a series of experiments carried on by Dr. Joule, of Manchester, between 1843 and 1849, the following important proposition was established:

The disappearance of a definite amount of mechanical energy is attended by the production of a definite amount of heat.

158 ELEMENTS OF PHYSICS.

The number of units of work necessary to produce one unit of heat is known as the *Mechanical Equivalent of Heat.* One method pursued by Dr. Joule in finding its value was that of measuring the heat produced in a vessel of water by agitating it with a set of paddles driven by a known weight falling through a known height. His conclusion was that *the quantity of heat which will raise one pound of water through* $1°$ *F. was equivalent to* 772 *foot-pounds of work or* 1390 *foot-pounds for* $1°$ *C.* If we employ metric units, the quantity of heat which will raise one kilogramme of water through $1°$ C. is equivalent to 424 kilogramme-metres of work. The elaborate experiments of Prof. Rowland in 1879 show that these values are 779 and 427.5 respectively.

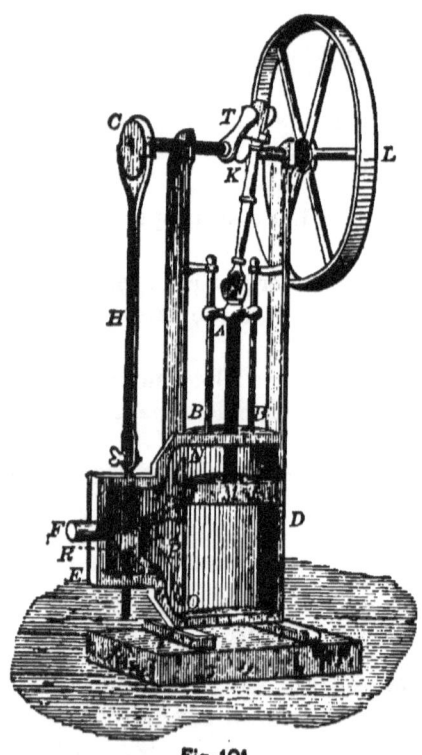

Fig. 101.

238. The Steam-Engine is a device for transforming the energy stored in steam into mechanical motion. In the more common of its many modern forms it consists of a strong cylinder in which a piston is made to move to and fro by applying the pressure of steam to its two faces alternately.

In Fig. 101 is shown an engine divested of many of the more complicated accessories designed to improve its efficiency. The cylinder is represented in section. The piston *M* moves to and fro in the cylinder *D* by virtue of the pressure of the steam supplied by the boiler through the tube *F*. In the *Steam-Chest E* works the *Slide-Valve R* which admits the steam alternately to the ends of the cylinder through *N* and *O*. When the valve is situated as shown, the steam passes into the upper end of the cylinder and drives the piston down. At the

Fig. 102.

same time the other end is connected with an exhaust-pipe shown at *P*, through which the steam either escapes into the air, as in *High-Pressure* or *Non-Condensing* engines, or into a large chamber, as in *Low-Pressure* or *Condensing* engines, where it is condensed to water, thereby reducing the pressure on that face of the piston. The slide-valve is moved by the rod *H*, connected to an *Eccentric C*, a wheel pivoted a little to one side of its centre, on the horizontal shaft *K*. This shaft receives its motion from the piston by means of the jointed rod *A*, and the

Crank T. The *Fly-Wheel L* serves the double office of belt-pulley and reservoir of energy. It is made with a heavy rim in order that when the piston is at the end of the cylinder, and the direction of motion must change, the energy stored up may be sufficient to carry the shaft beyond these *Dead Points* where the piston can again turn the shaft. It also serves to give uniformity of motion to the shaft, which would otherwise vary on account of the effective part of the force exerted on the crank not being constant, being considerable when the crank is at right angles to the connecting-rod, and diminishing to nothing when parallel to it.

In order that the piston-rod may always move in a straight line, and the piston maintain a steam-tight fit in the cylinder, the former is attached to a transverse bar, or cross-head *A*, which slides on two guide-bars *B*, *B*, firmly bolted to the framework of the engine, and adjusted accurately parallel to each other and to the piston-rod. In many large engines (Fig. 102) the cylinder is given a horizontal position.

CHAPTER V.

MAGNETISM AND ELECTRICITY.

I. MAGNETS. — POLARITY. — INDUCTION.

239. Natural Magnets. — There is found, widely distributed in nature, an iron ore, consisting of iron and oxygen, which sometimes possesses the property of attracting iron. This substance was probably first obtained near Magnesia, in Asia Minor, and hence the name *Magnet* is applied to it.

Fig. 103.

Exp. — Dip a piece of natural magnet into iron filings or small iron tacks. On withdrawing it, quite a large quantity will adhere in tufts to opposite parts of the magnet (Fig. 103).

Fig. 104.

Exp. — Make a stirrup out of wire, place in it the piece of natural magnet, and suspend it by an untwisted thread (Fig. 104) away from masses of iron. Carefully exclude all air-currents and allow the magnet to come to rest. Note the position, then disturb it slightly, and again let it come to rest. It will be found that it invariably returns to the same position, the line connecting the two parts to which the filings adhered in the preceding experiment lying north and south.

This property of the stone was early turned to account in navigation, and secured for it the name of *Lodestone* (leading-stone).

240. Artificial Magnets. — Exp. — Stroke a large darning-needle from end to end, and always in one way, with one of the ends of the lodestone. Dip it into iron filings and they will cling in tufts to it as they did to the lodestone. The needle has become a magnet.

Exp. — Use the needle of the last experiment to stroke another needle. This second needle also acquires magnetic properties, and the first one has apparently suffered no loss.

Artificial Magnets are those made from steel by the application of a lodestone or some other kind of magnetizing force. The process of making such a magnet is called *Magnetization*. The success of the operation depends largely on the quality and temper of the steel, the shape of the piece, and the mode of applying the magnetizing force. The principal forms of artificial magnets are the *Bar* and the *Horseshoe* (Fig. 105), so called from their shape.

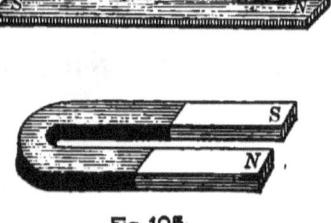
Fig. 105.

241. Polarity. — Exp. — Roll a bar magnet in iron filings. They will cling, in somewhat irregular tufts, near the ends, and but few, if any, near the middle (Fig. 106).

The experiment indicates that the attractive power of

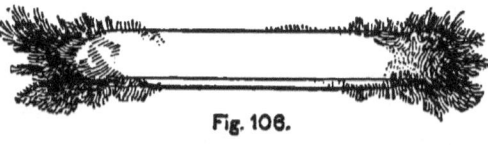
Fig. 106.

the magnet is concentrated in two opposite parts. These are called its *Poles*. The line joining these poles is the *Magnetic Axis*, and the line at right angles to the axis, marking the place of no attraction, is the *Equator* of the magnet.

242. Poles Distinguished. — Exp. — Lay a darning-needle on a piece of cork, floating in a glass vessel of water; note its position after it comes to rest (Fig. 107). Repeat the operation several times; it will generally be found that there is no uniformity in the direction it assumes when at rest. Now stroke the needle from end to end with one pole of a magnet and repeat the tests. It will then be seen that the needle always comes to rest lying nearly in a north and south line with the same end toward the north.

Fig. 107.

This fact has suggested that the name *North-Seeking Pole* be given to that pole of a magnet which tends to turn toward the north, and *South-Seeking Pole* to the opposite one.

243. Consequent Poles. — Exp. — Draw the temper of a knitting-needle slightly at two or three points. After stroking it with a strong magnet, roll it in iron filings. They will adhere in tufts along the needle as well as at the ends, showing the existence of several poles.

It has been found impossible to make a magnet with but one pole, two poles being the least number possible. Sometimes, however, through the lack of uniformity in the steel, or through defects in the method of magnetization, other intermediate poles are formed, called *Consequent Poles*.

244. Action between Magnets. — Exp. — Suspend in succession, after the manner shown in Fig. 104, two similar bar magnets and determine which are their N-seeking poles. Now leaving one of them suspended, bring successively to its N-seeking pole the poles of the other magnet. In like manner, present them to the S-seeking pole. The results obtained are conveniently expressed in the following law:

Like poles repel each other, and unlike poles attract.

245. Compound Magnet. — Exp. — Dip one pole of a bar magnet into a dish of small iron tacks and observe the quantity adhering on withdrawing it. Now slide the opposite pole of a second bar magnet over it, when most or all of the tacks will fall off, showing that the second magnet has nearly or wholly neutralized the attractive power of the first one. Repeat the experiment with the two magnets placed side by side, like poles adjacent. The number of tacks lifted will be considerably greater.

Two or more magnets so grouped as to make their like poles coincident form a *Compound Magnet*.

246. A Magnetic Substance is one which is attracted by a magnet or is capable of being magnetized. Faraday proved that almost all substances are influenced by powerful magnets. Ordinary magnets, however, affect noticeably but few substances besides the compounds of iron. Cobalt, nickel, and liquid oxygen are about the only ones worthy of mention. Steel is said to lose its magnetic qualities when alloyed with 20 per cent. of manganese. Some substances, like bismuth and antimony, are slightly repelled by strong magnets, and are called *Diamagnetic*.

247. Magnetic Transparency. — Exp. — Cover the pole of a strong bar magnet with a thin plate of glass. Bring the face of the plate opposite the pole in contact with a pile of iron tacks. A number will be found to adhere, showing that the attraction takes place through glass. In like manner, try thin plates of mica, wood, paper, zinc, copper, and iron. No perceptible difference will be seen except in the case of the iron, where the number of tacks lifted will be considerably less.

Magnetic force acts freely through all substances, except those classed as *Magnetic*. Soft iron serves as a screen to magnetism, more or less completely, depending on the mass used and the power of the magnet. Watches are now protected from magnetism by using an inside case of soft iron.

248. Induced Magnetism. — Exp. — Hold one end of a short rod of soft iron near one pole of a strong bar magnet, and while in this position dip the other end into iron filings. They adhere to it as to a magnet, but fall off on removing the magnet.

Magnetism produced in magnetic substances by the influence of a magnet is said to be *Induced*.

249. Polarity of the Iron Bar. — Exp. — Support a strong horseshoe magnet in a vertical plane, with its poles uppermost, and the line joining them horizontal (Fig. 108). Suspend by a thread a short rod of soft iron so that it hangs horizontally above and near the poles of the magnet. Now bring near one end of this rod a bar magnet, so that its pole is opposite in name to that of the vertical magnet. The repulsion of the rod indicates that its polarity is the same as that of the bar magnet, and hence the reverse of that of the horseshoe magnet.

Hence, it appears that when a magnet is brought near a piece of iron it magnetizes it, and that the attraction exhibited is that between unlike poles. The inductive action can take place through a series of iron rods, a phenomenon seen in the attraction of a bunch of filings or tacks.

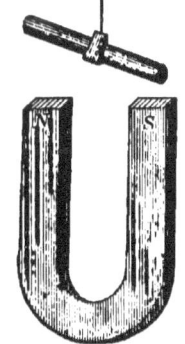

Fig. 108.

250. Inductive Action of Magnets on Magnets. — Exp. — Support a bar magnet in a horizontal position and hold up to one pole a quantity of iron tacks, noting how many adhere. Place a second magnet beneath this one, so that opposite poles are adjacent and about six or seven centimetres apart. Again test the upper magnet with the tacks. Its power will be greatly increased. Now reverse the position of the lower magnet, placing like poles adjacent; the power will be lessened.

In the first case, the S-seeking pole of the second magnet induced a stronger N-seeking pole in the first one, as shown by the increased number of tacks lifted. In the second case, the effect was the reverse.

II. NATURE OF MAGNETISM.

251. Magnetism a Molecular Phenomenon. — Exp. — Magnetize a darning-needle, then heat it red hot and test it for magnetism. It will be found to have lost the power of attracting filings.

Exp. — Magnetize a knitting-needle and find by averaging several trials how many tacks can be lifted by it. Now hold one end firmly against the edge of the table and, by plucking the free end, cause the needle to vibrate vigorously for a few seconds. On testing the power of the magnet to pick up tacks it will be found considerably lessened.

Exp. — Take a piece of iron wire, about 30 cm. long and 1.5 mm. diameter, and carefully anneal it. Bend it to the form shown in Fig. 109. Stroke it carefully several times with a strong magnet. It will be a magnet. (How shown?) Now hold it by the turned-up ends and give the wire a sudden twist. On retesting it nearly all magnetism will be gone.

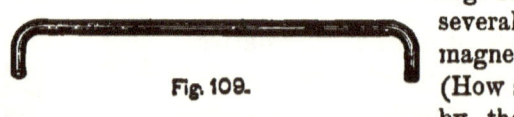

Fig. 109.

In each of the preceding experiments the molecular arrangement has been disturbed; and it is interesting to note that in each the magnetism also has been disturbed.

The conclusion is that, whatever magnetism is eventually proved to be, it will in some way be connected with the molecular arrangement of the substance.

252. Further Evidence. — Exp. — Magnetize a knitting-needle; notice that it has two poles, one at each end, the centre being neutral. Break it at the neutral point; the pieces will be found to have two poles, two new ones having been formed at the point that was formerly neutral. If these pieces be in turn broken, their parts will be magnets similar in character to the original.

There seems to be no limit to the extent to which this process may be carried, indicating that possibly if carried as far as the molecules, they may prove to be magnets.

Exp. — Fill a slender glass tube nearly full of steel filings, closing the ends with cork. Stroke the tube from end to end with a strong magnet; it acquires magnetic properties. Shake up the filings thoroughly; all polarity is lost.

An examination of each steel particle will show that it is a magnet. The loss of polarity is evidently due to the neutralization of the actions of many little magnets through their indiscriminate arrangement. The polarity would probably be restored if the particles could be restored to their original positions. The experiment strongly supports the theory that each particle of a magnetic substance is a magnet.

It is worthy of notice that magnetization is facilitated by jarring the substance, or by heating it and then cooling it while under the magnetizing influence.

253. Retentivity. — Exp. — Prepare three bars of the same size, one each of soft iron, soft steel, and hard steel. Successively dip one end of each into iron filings, and bring a strong magnet in contact with the other end. On withdrawal the greatest quantity of filings will adhere to the soft iron and the least to the hard steel.

On removing the magnet most of the filings drop from the iron, and the hard steel holds the most.

The difference exhibited by these substances is due to what is called the *Retentivity*, or the ability to retain magnetism.

254. The Physical Theory of Magnetism. — The foregoing experiments have suggested the following theory of magnetism:

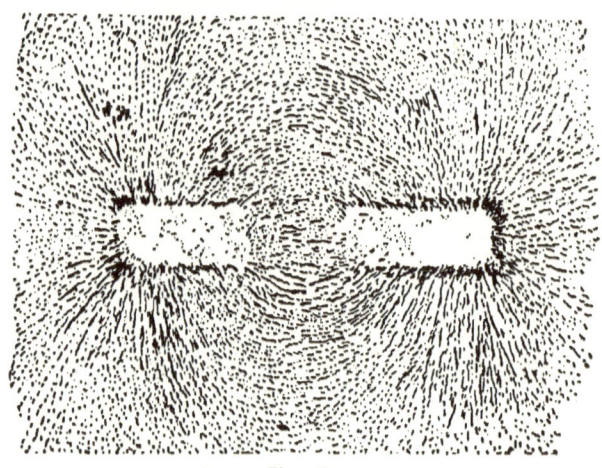

Fig. 110.

Each molecule of the magnetic substance is a magnet. In an unmagnetized bar the arrangement is such that externally the polarity of the different molecules is neutralized. These molecules resist displacement from their normal positions. A rearrangement is, however, more or less completely accomplished when the substance is brought under the influence of a magnet. When all the molecules have their like poles pointing the same way the bar is magnetized to saturation.

III. THE MAGNETIC FIELD.

255. A Magnetic Field is the space surrounding a magnet within which magnetic substances are influenced. Faraday's method of studying the distribution of magnetism in such a field is illustrated in the following experiment:

Place a sheet of paper or glass over the magnet and then sift iron filings evenly over it from a muslin bag, tapping the paper gently to facilitate the movement of the filings. They arrange themselves in

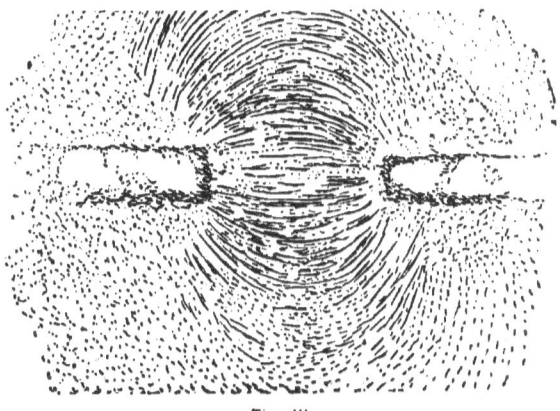

Fig III.

curved lines, diverging from one pole of the magnet and meeting again at the opposite pole.

These lines are called *Lines of Force*, or *Lines of Magnetic Induction*. Each particle of iron on falling on the paper becomes a magnet through induction; hence the lines mapped out by the filings are the lines along which magnetic induction takes place. Fig. 110 shows the lines of force in one of the planes which passes through the magnet. They are to be considered as extending outward from the N-seeking pole of the magnet, curving round through the air to the S-seeking pole, and completing the circuit back through the magnet. Fig. 111 shows the field between the unlike poles of two magnets when near each other; Fig. 112 between the poles of a horseshoe

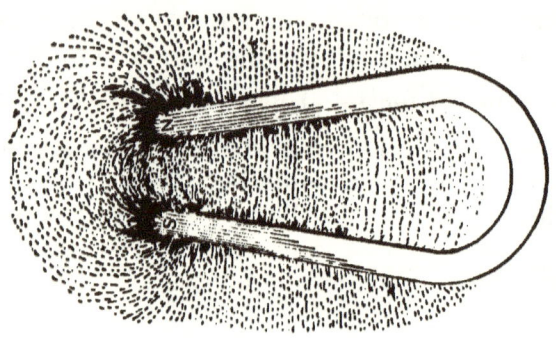

Fig. 112.

magnet; and Fig. 113 between the like poles of two bar magnets. In the last case the lines curve away from one another as if repelled. Lines of magnetic force are

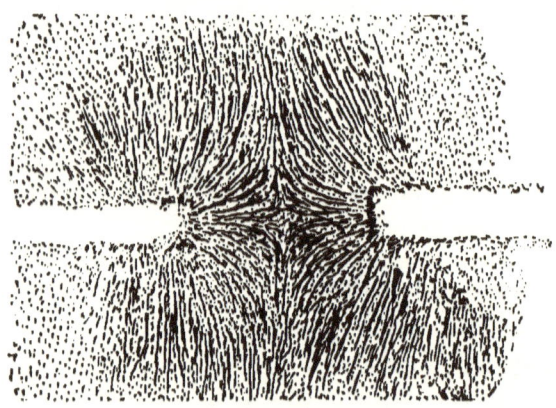

Fig. 113.

always to be regarded as possessing elasticity or resiliency, and to be under tension. They map out the lines of magnetic strain in the ether of a magnetic field. When for any reason these lines are distorted, their resiliency always causes them to tend to recover from the distortion or to react against it. Hence it is easy to see that Fig. 111 is a picture of the attraction exhibited by poles of unlike sign, while Fig. 113 is one of repulsion between poles of the same sign.

256. Direction of Lines of Force. — The direction of a line of force at any point is that of the rectilinear tangent to the curve at that point; it is the same as that toward which a free N-seeking pole is urged. If a magnetic needle, free to move about a vertical axis, is brought near the N-seeking pole of a magnet, as has been shown, its N-seeking pole is repelled. Hence, if an observer stands with his back to the N-seeking pole of a magnet, he is looking in the direction of the lines of force which pass out from that pole.

(Queries: What is the direction of the lines of force opposite the middle of a bar magnet? What is the direction of the lines of force between the poles of a horseshoe magnet?)

257. Permeability. — Experiment shows that if a piece of iron is placed in a magnetic field, the lines of force are concentrated by it. This property possessed by iron of increasing the number of lines of force when placed in a magnetic field is known by the term *Permeability*. Iron offers superior facilities for the formation of lines of force, and this fact explains the action of magnetic screens (247). In the case of the watch shield, the lines of force pass through the iron and not across it, and the watch is thus protected from magnetism because the lines of force do not enter it.

IV. TERRESTRIAL MAGNETISM.

258. The Earth a Magnet. — **Exp.** — Place a long bar magnet on the table and suspend over it a magnetic needle mounted so as to turn readily about a horizontal axis. When over the N-seeking pole, the needle will be vertical, with its S-seeking pole down; when over the middle, it is horizontal; and when over the S-seeking pole, the needle is again vertical, but reversed in position.

In like manner the earth acts toward such a needle when moved over its surface from pole to pole. At a point in Boothia Felix, west of Baffin's Bay, the needle is nearly vertical, with its N-seeking pole down; at points successively farther south, it is inclined to the vertical, becoming horizontal near the earth's equator, and again gradually inclining toward the vertical, with its S-seeking pole down as it nears the south magnetic pole of the earth. If a bar magnet, about half the length of the earth's diameter, were thrust through the earth's centre, making an angle of about $20°$ with its axis, it would account for many of the phenomena of terrestrial magnetism.

259. Earth's Induction. — **Exp.** — Procure a thoroughly annealed iron bar 75 cm. long, showing little or no polarity when tested with a magnetic needle while the bar is supported horizontally in an east-and-west line. Hold this bar in a meridian plane, but with its north end dipping down some $70°$ below the horizontal. Tap the bar with a hammer and then test for polarity. The lower end will be found strongly N-seeking, and the upper end S-seeking. If the bar is turned end for end, and again tapped with a hammer, the lower end again becomes N-seeking.

The experiment illustrates the inductive action of the earth. If we examine any iron object which has remained undisturbed for some time, as the stove or the supporting columns of a building, we shall find that they are polarized similarly to the bar of the above experiment. The inductive action of the earth undoubtedly accounts for the existence of natural magnets.

260. Magnetic Dip. — Exp. — Thrust through a cork an unmagnetized knitting-needle, and at right angles to this two short pieces (Fig. 114). Support the apparatus on the edges of two glass tumblers, with the axis in an east-and-west line, and the needle adjusted so as to rest horizontally. Now magnetize the needle, being careful not to displace the cork. It will no longer assume a horizontal position, the N-seeking pole dipping down as if it had become heavier.

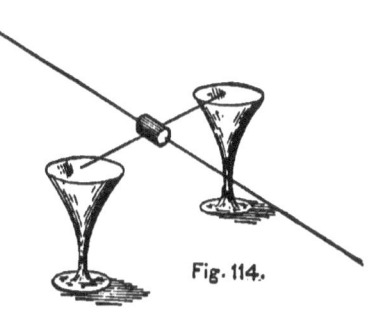

Fig. 114.

The angle made by this needle with the horizontal plane is called the *Inclination* or *Dip* of the needle. A magnetic needle mounted so as to move freely in a vertical plane, and provided with a graduated arc for measuring the inclination is called a *Dipping Needle* (Fig. 115).

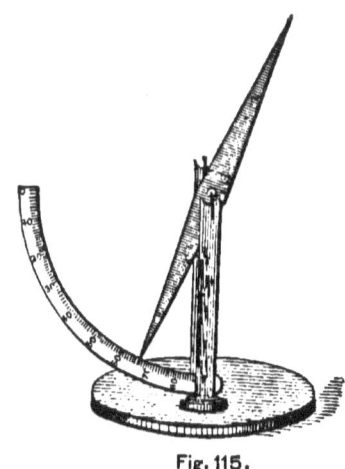

Fig. 115.

The magnetic poles of the earth are points where the dip is 90°; the dip at the magnetic equator is 0°. Lines on the earth's surface, passing through points of equal dip, are called *Isoclinic Lines;* they are irregular in character, though resembling in some measure the parallels of latitude.

261. Declination. — The magnetic poles do not agree

with the terrestrial poles, and consequently in most places the direction of the magnetic needle will not be that of the meridian of the place. The direction of the magnetic needle at any place is that of the magnetic meridian of that place. *The Declination* of the needle is the angle between the magnetic and the geographical meridian.

262. The Line of no Declination passes through those places where the needle points true north. Such a line, in 1890, ran from the north magnetic pole (situated west of Baffin's Bay) across Lake Superior, passed near Ann Arbor, Mich., Columbus, Ohio, through West Virginia and South Carolina, and left the mainland near Charleston on its way to the south magnetic pole. The returning line through the eastern hemisphere is quite irregular in direction. At places east of this line the needle points west of north, and west of the line it points east of north. Lines passing through points of equal declination are called *Isogonic Lines*.

V. STATIC ELECTRICITY.

263. Electrical Attraction. — **Exp.** — Cut a number of small balls out of the pith of common elder. Place them in a pile on the table and touch them with a rod of sealing-wax. Notice that the rod does not affect the balls in the least. Now rub the rod with dry flannel and again bring it up to the pile of balls. They will be alternately attracted and repelled.

Rods of glass, shellac, sulphur, very dry wood, ebonite, etc., may be substituted for the sealing-wax, and a collection of any light objects for the pith-balls.

Bodies which exhibit the power of attracting light bodies after being rubbed are said to be *Electrified*. *Elec-*

trification may be brought about in a variety of ways, as will appear in the course of this chapter.

264. Attraction Mutual. — Exp. — Prepare a glass tube of about 2 cm. diameter and 40 cm. long. Remove all sharp corners by fusion in the flame of a blow-pipe. Electrify the tube by friction with a piece of silk, and hold it near the end of a long wooden rod resting in a wire stirrup suspended by a silk thread (Fig. 116). The suspended rod is attracted. Now, replace the rod by the electrified tube. On holding the rod near the rubbed end of the glass tube, the latter moves as if attracted by the former.

Fig. 116.

The experiment teaches that each body attracts the other; that is, that *the action is mutual*.

265. Electrical Repulsion. — Exp. — Suspend several pith-balls by fine linen threads from a glass rod, and touch them with an electrified glass tube (Fig. 117). At first they are attracted, but soon fly away from the tube and from one another. On removing the tube the balls no longer hang side by side, but keep apart for some little time. If we bring the hand near the balls they will move toward it as if attracted, showing that the balls are electrified.

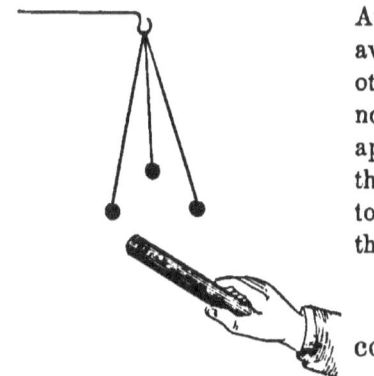

Fig. 117.

It thus appears that bodies become electrified by coming in contact with electrified bodies, and also that electrification may manifest itself by repulsion as well as by attraction.

266. Two Kinds of Electrification. — Exp. — Rub a glass tube with silk and suspend it as in Fig. 116. Excite a second glass tube and hold it near one end of the suspended one. The suspended tube is repelled. Bring near the suspended tube a rod of sealing-wax excited by friction with flannel. The suspended tube is now attracted. Repeat these tests with an electrified rod of sealing-wax in the stirrup instead of the glass tube. The electrified sealing-wax will repel the electrified sealing-wax, but there will be attraction between the sealing-wax and the glass tube.

The experiment indicates that there are *two kinds of electrification: one* developed by rubbing glass with silk, and the *other* by rubbing sealing-wax with flannel. In the former case the body is said to be *positively* electrified; in the latter case *negatively* electrified.

Fig. 118.

267. Action of Electrified Bodies toward Each Other. — It was seen in the last experiment that there was repulsion between the electrified glass tubes, and that the electrified sealing-wax attracted the electrified glass. These facts are expressed by the following law:

Electrical charges of like sign repel each other; electrical charges of unlike sign attract.

MAGNETISM AND ELECTRICITY.   177

268. The Electroscope, as the name implies, is an instrument for detecting electrical charges. The most common form is the *Gold-leaf Electroscope*. It consists of a glass flask, through the cork of which passes a brass rod terminating in a ball or disk (Fig. 118) on the outside, and two strips of thin metal foil, preferably aluminium, on the inside, hanging parallel and close together. If an electrified object is brought in contact with the rod, the metal strips become similarly charged, and hence are mutually repelled.

269. Use of the Electroscope. — In order to determine the kind of electricity with which a body is charged, transfer some of the charge to the ball of the electroscope by means of a *Proof-plane* (Fig. 119), which is usually a small metal disk

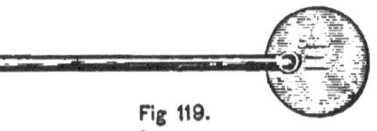

Fig 119.

cemented to the end of a glass, ebonite or shellac rod. To charge the electroscope, slide the metal disk of the proof-plane along the surface to be tested, and then touch it to the knob of the electroscope. Now excite a dry glass tube by friction with silk. Charge the proof-plane from the tube, and then, without delay, transfer this charge to the electroscope. If the leaves diverge farther, the body in question was positively charged; if the leaves collapse, its charge was probably negative. Since the leaves would collapse if the proof-plane were either neutral or less highly charged with the same kind of electricity, an increased divergence of the leaves is the only sure test. Hence, whenever the leaves collapse, repeat the test, using a rod of sealing-wax and a rubber of flannel in place of the silk and the glass tube.

270. Simultaneous Development of the two Electricities. — Exp. — Fit to the end of a rod of sealing-wax a cap of flannel, three or four inches long, with a silk cord attached to the end by which it can be drawn off (Fig. 120). Electrify the rod by turning it around inside of the cap, and then touch it to the knob of the electroscope. No divergence will be observed. By the aid of the cord remove the flannel cap, and present it to the positively charged electroscope. The increased divergence shows that the cap is positively charged. In like manner, if we test the rod of sealing-wax, it will be found to be negatively charged.

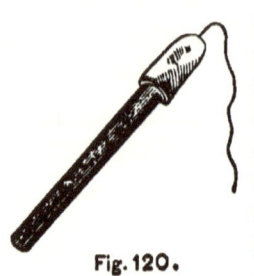

Fig. 120.

The experiment shows (1) *that one kind of electricity is not developed without a development of the other;* and (2) *that the two kinds of electricity are produced in equal quantities*, as was shown by the fact that the quantity in the rubber exactly neutralized that on the rod when the two were in contact. The two charges behave like equal positive and negative quantities.

271. Conduction. — Exp. — Support a smooth metallic button on a rod of sealing-wax. Connect it to the knob of the electroscope by a fine copper wire, 50 to 100 cm. long. Touch the button with an electrified glass tube. The divergence of the leaves indicates that they are electrified. If we repeat the experiment, using a silk thread in place of the wire, no effect is produced on the leaves.

The communication of electricity to a body through a second one is called *Conduction*, the communicating body serving as a *Conductor*. As seen in the above experiment, bodies differ widely in the readiness with which they conduct electricity. Those offering very little resistance to its passage are called *Good Conductors*, those which offer such great resistance that practically no passage occurs are called *Non-Conductors*, *Insulators*, or *Dielectrics*. Between these two extremes we find a great many bodies varying in their conductivity, so that it is not possible to draw a sharp line between the two classes. The following classification is only relative:

Good Conductors: Metals and carbon.

Semi-Conductors and Bad Insulators: Water, aqueous solutions, moist bodies, wood, cotton, hemp, liquid acids, rarefied gases.

Good Insulators: Paraffin, turpentine, silk, sealing-wax, India-rubber, gutta-percha, dry glass, porcelain, mica, shellac, air at ordinary pressures, liquid oxygen.

Conductivity in bodies is affected by temperature; some insulators, like glass, become good conductors when heated.

272. **Probable Nature of Electrification.** — It was suggested by Faraday, and a multitude of facts tend to confirm his views, that electrification is a strained condition of a body communicated to it by the ether which surrounds it and pervades it. Conductors differ from insulators in this: in the former, the molecular mobility is such that this state of strain is continually giving way, whereas in the latter considerable distortion is possible before the molecular structure yields to the strain. The phenomena of attraction and repulsion exhibited by electrified bodies are due to the attempt of the strained ether in and around the bodies to return to its normal condition. In producing electrification, work is done in distorting the medium; hence electrification is a form of potential energy. Electricity, however, is not energy.

VI. INDUCTION.

273. Electrification by Induction. — **Exp.** — Excite a glass tube by friction with silk. Bring it gradually near the ball of the electroscope. The leaves begin to diverge when the tube is some distance from the knob, and the amount of divergence increases as the tube is brought nearer. On removing the tube the leaves collapse.

It is evident, since the leaves do not remain apart, that there has been no transfer of electricity from the tube to the electroscope. The electrified condition, produced in the electroscope when the electrified body is brought near it, is due to what is called *Electrostatic Induction*. Why such an effect should occur is easily understood when we recall Faraday's views of electrification, that it is a distortion of the ether about the body. Evidently, then, any body placed within this *Electrical Field* should be electrified.

274. Charging a Body by Induction. — **Exp.** — Support a smooth metallic ball on a dry plate of glass. Connect it with the knob of the electroscope by means of a metallic wire, the ends of which are bent into a loop and smoothly soldered. The ball and the electroscope now form one continuous conductor. Bring near the ball an electrified glass tube; the leaves of the electroscope diverge. Before removing the excited tube, remove the wire, handling it with some non-conductor. The electroscope remains charged. On testing it, it will be found to be positive. A similar test made of the ball will show that it is negatively charged. Repeat the experiment without removing the connecting wire. No signs of electrification exist on removing the excited tube.

Hence, we learn two things: (1) *When an electrified body is brought near an object it induces the opposite kind of electricity on the side next it and the same kind on the remote side;* (2) *the two kinds of electricity are developed in equal quantities.*

MAGNETISM AND ELECTRICITY. 181

275. Attraction explained. — Since an electrified body induces the opposite kind of electricity on the near side of a neighboring object, an attraction will, in consequence, exist between them. If the neighboring object were a non-conductor, the electrical separation would be prevented, and no attraction would take place. This may be proved by suspending a small ball of sealing-wax by a silk thread and bringing near it an electrified glass tube.

276. Charging an Electroscope by Induction. — Exp.— Hold one finger on the ball of the electroscope and bring near it an electrified glass tube. Remove the finger before taking away the tube and the electroscope will be charged. Explain. What kind of electricity will then be in the electroscope? How can you modify the intensity of the charge?

277. Inductive Capacity. — Exp.— Suspend an electrified brass ball above an electroscope, and distant far enough to produce only a slight effect on the leaves. Take a cake of paraffin or sulphur whose thickness is a little less than the distance between the ball and the electroscope, pass a gas flame over its surface to remove all traces of electrification, and insert it between the ball and the electroscope knob, being careful not to touch either. The leaves of the electroscope will diverge further, as if the ball were brought nearer to the knob.

The experiment shows that *induction depends on the nature of the intervening medium*. The property which bodies have of transmitting electrical induction is called *Specific Inductive Capacity*, and the bodies themselves are called *Dielectrics*. All non-conductors or insulators are dielectrics; but equally good insulators have different specific inductive capacities.

VII. ELECTRICAL DISTRIBUTION.

278. Effect of Material.— Exp.— Rub one end of a glass tube with silk. On testing it electrification will be found confined to the end rubbed. Touch one end of an insulated metallic conductor with an electrified substance. On testing it all parts will be found electrified.

Hence, it appears that *the distribution of electricity over a surface is dependent on the electrical conductivity of the material.*

279. The Charge on the Outside of a Conductor. — Exp. — Support a metallic cylindrical vessel of about one litre capacity on an insulated support (Fig. 121). One free from sharp edges should be selected. Electrify strongly and test in succession both the inner and the outer surface, using a proof-plane to convey the charge to the electroscope. It will be found that the inner surface gives no sign of electrification.[1]

Fig. 121.

Hence, it appears that *the electrical charge of a conductor is confined to its outer surface.*

Fig. 122.

280. Effect of Shape. — Exp. — Charge electrically an insulated egg-shaped conductor (Fig. 122). Charge the proof-plane from it by placing it against the large end, and then convey the charge to the electroscope. Notice the amount of separation of the leaves. In like manner test the small end of the conductor. A greater divergence of the leaves will be observed in the latter case.

The distribution of the charge is, therefore, affected by the shape of the conductor, the electrical

---

[1] Evidence of electrification may be found on the inside of the vessel near the mouth.

density being greater the greater the curvature. By *Electrical Density* is meant the quantity of electricity on unit area of the surface of the conductor. The last experiment shows that the electrical density is greatest at the small end of the conductor.

281. Effect of Area. — Exp. — Employing an electroscope provided with a disk instead of a ball, place on it a chain, and charge the electroscope by induction so that the leaves diverge widely. Now lift the chain by a dry glass rod, thereby increasing the surface. The leaves of the electroscope will slowly collapse. On lowering the chain they will again diverge.

Hence, *with the same charge, the electrical density increases as the surface diminishes.*

282. Action of Points. — Exp. — Cement a pin at the middle to the end of a glass tube. Charge the electroscope and then place the head of the pin in contact with the knob. Observe that the electroscope is rapidly discharged.

From this experiment it is apparent that *a conductor provided with points cannot retain an electric charge*. This conclusion might be drawn from Art. 280. If, by increasing the curvature of the end, the density of the charge becomes very great, the charge communicated to the adjacent air-particles is also correspondingly great, thus facilitating the discharge. These particles are charged with electricity of the same sign as the conductor and are repelled. Their places are then taken by other particles, which are in turn charged and repelled. The air current thus produced is called an *Electric Wind*.

283. Applications. — Since the charge is confined to the outer surface of a body, it follows that if a hollow conductor envelop the electroscope, the latter will not be affected by any charge on the envelope or by any charge external to it. Such an electric screen can be made out of fine wire-cloth.

In placing the proof-plane on any conductor to test its electrical condition, care must be taken to lay it as flat as possible, in order that the electrical distribution on the conductor may not be disturbed by any change of shape due to the proof-plane.

On account of the action of points, electrical apparatus should have the parts smoothly rounded and free from dust.

EXERCISES.

1. Why will not an electrified body remain charged for an indefinite length of time?

2. If a positively charged body be suspended by a silk thread within a hollow insulated conductor what will be the electrical condition of the inner surface? What will be the condition of the outer surface?

3. Why must a metal rod be separated from the hand by some such substance as rubber in order to electrify it by friction.

4. Why must all electrical appliances be kept free from moisture?

5. Account for the fact that bringing the point of an uninsulated needle near the knob of the electroscope discharges it.

VIII. ELECTRICAL CAPACITY.

284. Potential. — When an electrified body is brought in contact with an insulated conductor, an electric charge is given to the entire surface of the latter. Since electrification is energy in the potential form, the flow of electricity from one body to another, or from one part of a conductor to another part, is determined by what is known as the relative *Potential* of the two bodies or parts of the same body. It is analogous to temperature. As heat flows from places of higher to places of lower temperature, tending to equalize the difference of temperatures, so electricity flows from bodies of higher potential to those of lower potential. The flow continues till an equalization of potential is reached. This flow is known as an *Electric Current*.

The *Difference of Potential* between two conductors is the work required to transfer unit quantity of electricity from one conductor to the other.

285. Zero Potential. — For purposes of comparison and measurement the potential of the earth is assumed to be zero. Although different parts of the earth's surface are not at the same potential or electrical level, yet the potential of the earth is a convenient standard of reference. A body positively charged is at a higher potential than the earth, and one charged negatively is at a lower potential. If a conductor of positive potential be connected to the earth by an electrical conductor, positive electricity will flow to the earth. If the conductor is at a negative potential, the flow is in the other direction.

286. Difference of Potential made Visible. — **Exp.** — Charge an electroscope positively, producing a small separation of the leaves. In like manner charge a second one, producing a wide separation of the leaves. The latter is charged to the higher potential. Connect the knobs by a wire, handling it by means of a non-conductor. The divergence of the first electroscope increases and the second one decreases, showing that electricity flows from the place of higher potential to that of lower.

287. The Electric Spark. — **Exp.** — Turn the handle of an electrical machine (298) a few times, and hold the knuckle near the metallic conductor. A bright spark will pass between them. Observe that a longer spark can be got by giving the machine several turns before bringing the hand near the conductor.

The spark is produced because the difference of potential, or the electric strain, is so great that the resisting air gives way, and a transfer of electricity takes place between the machine and the hand. The potential energy of the electric charge is converted into heat.

288. Electrical Capacity. — Exp. — Suspend a small, smooth metallic ball by a silk thread. Charge the gold-leaf electroscope till the leaves diverge widely. Bring the small ball in contact with the knob of the electroscope; the leaves will partially collapse, showing that the potential has been lowered.

The *Electrical Capacity* of a conductor means the quantity of electricity necessary to raise it from zero to unit potential. The above experiment shows that *it is dependent upon the extent of surface of the conductor.*

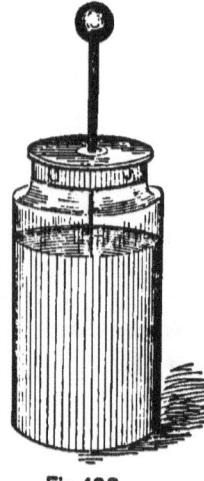

Fig. 123.

289. Condensers are devices for holding large quantities of electricity. They increase the electrical density without increase of potential. Two conductors separated by a dielectric constitute a condenser.

290. The Leyden Jar is the most common and convenient form of condenser. It consists of a glass jar coated part way up, on both the inside and the outside, with tin-foil (Fig. 123). Through the wooden or ebonite stopper passes a brass rod, terminating on the outside in a ball or disk and communicating with the inner surface of the jar by a metallic chain. The inner foil represents the collecting surface, and the glass the dielectric separating the two conductors.

291. Charging and Discharging Jars. — To charge a Leyden jar connect the outer surface with the earth, either by a metallic conductor or by holding the jar in the hand. Place the ball in contact with the source of electricity, as, for example, the conductor of an electrical machine. To discharge a Leyden jar bend a wire into the form of the letter V. With one end of the wire touching the *outer* surface of the jar (Fig. 124), bring the other around till it touches the ball, and the discharge will take place.

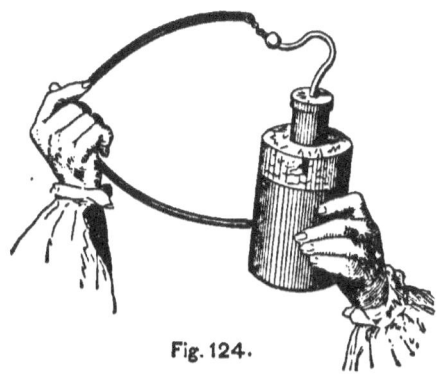

Fig. 124.

292. Action of Jar explained. — When a charge of positive electricity is communicated to the inner surface it acts inductively through the glass, making the outer sur-

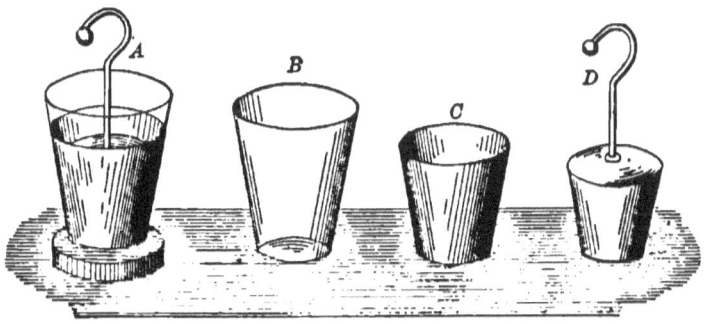

Fig. 125.

face negative. These two charges act inductively on each other through the glass and are said to be "bound," in distinction from that condition where a conductor is charged with electricity and is at some distance from any conductor, in which case the whole charge is "free." If the outer surface of the jar is connected with the earth, the electrical capacity of the jar is largely increased.

293. Seat of Charge.— Exp.—Charge a Leyden jar made with movable metallic coatings (Fig. 125). Lift out the inner coating D by means of a glass tube. Then remove the outer coating C from the glass vessel B. The coatings exhibit no sign of electrification. Bring the glass vessel near a pile of pith balls; they will be attracted to it, showing that the glass is electrified. Now build up the jar by putting the parts together; the jar will still be found to be highly electrified and may be discharged in the usual way.

This experiment, due to Franklin, shows that the electrification is a phenomenon of the glass. Faraday proved that during the act of charging the jar the glass is strained, the office of the conductors being to facilitate the release from strain. This is supported by the facts that thin jars can be broken by over-charging; that a jar enlarges on charging; that on heating a jar its charge disappears; and that on charging a jar heavily and then discharging in the usual way a second charge accumulates after a few minutes, the time being lessened by tapping the jar, as if the glass were strained or distorted to so great a degree that, like a twisted glass fibre, it does not return at once to its nor-

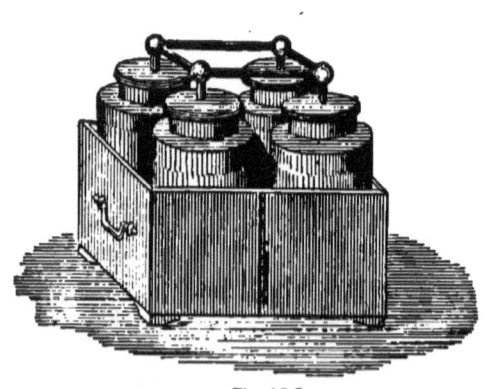

Fig. 126.

mal state when released. This second charge is called the *Residual Charge.*

294. Battery of Leyden Jars. — Since the capacity of a condenser is proportional to its surface, it follows that a greater quantity of electricity can be stored by connecting together the inner surfaces of two or more Leyden jars by conductors, and at the same time joining their outer surfaces. Fig. 126 illustrates such a battery.

EXERCISES.

1. Stand a charged Leyden jar on a cake of paraffin. Touch the ball; it is not discharged. Why?
2. Connect the inner surface of an uncharged Leyden jar to an electroscope, and insulate the outer surface from the earth. Notice that a very small quantity of electricity causes a violent separation of the leaves. Discharge and connect the outer coating to the earth. A much greater quantity is needed to affect the electroscope to the same extent as before. Explain.
3. If a fine wire two or three feet long connect the knob of the electroscope to a proof-plane, and the proof-plane be placed in contact with different parts of the egg-shaped conductor (Fig. 122) when charged, no difference in the divergence of the leaves will be noticed. Explain. Now remove the wire, discharge the electroscope, then test different parts of the egg-shaped conductor, and explain why one proof-plane charge from the point affects the electroscope more than one charge from any other part. How do you reconcile this result with the first one?

IX. ELECTRICAL MACHINES.

295. The Electrical Machine. — The quantity of electricity which can be obtained by rubbing a glass tube or a rod of sealing-wax is very small. When larger quantities are needed, larger surfaces must be used and some more convenient method must be devised to excite them. *The Electrical Machine* is such a device. It consists essentially of two parts, one for producing, the other for collecting, the electricity.

296. The Electrophorus, invented by Volta in 1775, consists of a bed of resinous material, or a disk of vulcanite, resting on a metallic plate (Fig. 127), and a metallic disk provided with an insulating handle.

Fig. 127.

To use the instrument, rub the bed with a warm woollen cloth or strike with a cat's skin. Rest the metallic disk upon it and touch the upper surface momentarily with the finger. Lift the disk by the insulating handle; it will be found highly electrified. On discharging it, it may be recharged many times before the bed loses any perceptible portion of its charge.

297. Explanation. — If the bed be tested, it will be found to be negatively charged. Since its surface is uneven and the material is non-conducting, the metallic disk touches it at only a few points, so that scarcely any of the electricity passes from the bed to the disk, unless they are left in contact for a long time. The two disks, with a very thin layer of air between, form a condenser of great capacity. The electrified bed, acting by induction on the disk, develops positive electricity on the lower side and negative on the upper. The negative is removed on touching the disk with the finger, leaving it positively charged. On raising the disk, the charge, being no longer "bound" by that of the bed, distributes itself over the surface.

298. The Holtz Machine is a representative of a class of machines depending on the employment of a small

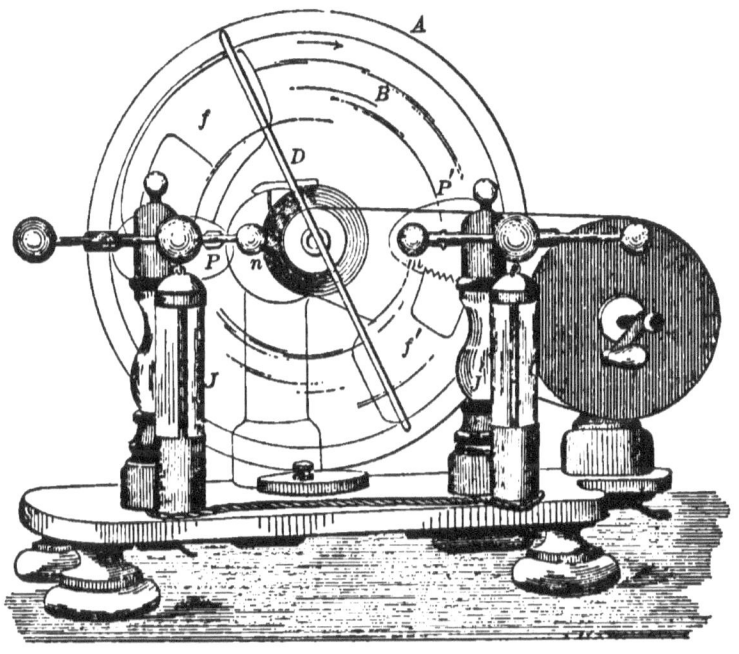

Fig. 128.

initial charge which acts inductively on the conductors of the machine, producing other charges which are conveyed by the moving parts to other points, there to increase the initial charge or electrify a suitable conductor. In its latest form, it consists of two vertical plates, A and B (Fig. 128), placed about half an inch apart, the former stationary, and the latter arranged to rotate on an insulated axis. The stationary plate is usually about two inches greater in diameter than the other, and has two openings, P and P', called *Windows*, cut at opposite points on a nearly horizontal diameter. Two varnished pieces of paper, f and f', called *Armatures*, are cemented to the plate, one above the left-hand window, and the other below the right. That portion of the paper armatures next to the windows is usually of gilt paper, and cut with teeth projecting part way into the openings. On the front side of the moving plate, and opposite the serrated edges of the armatures, are insulated conductors with sharp points nearly touching the plate. These conductors are called *Combs*. At right angles to these are two brass rods connecting them with two large brass balls, through which slide two rods, terminating in the balls m and n, which form the *positive* and *negative conductors* of the machine, as well as a discharging apparatus. Also on the front of the moving plate is a diagonal conductor D, the inner faces of the parts extending over the paper armatures being furnished with sharp points. The inner surfaces of small Leyden jars, J and J', are connected to the system of conductors to increase their capacity.

To work the machine, place m and n in contact, and charge one of the armatures, using a Leyden jar or electrophorus for the purpose. Rotate the plate steadily and rapidly, the direction of motion being toward the teeth on the armature. After a few turns, slowly separate m and n; a shower of bright sparks will pass between them and greater effort will be necessary to drive the machine.

299. The Action of the Machine is in general as follows: Suppose m and n in contact and f' positively charged. This armature acting on the conductor opposite induces negative electricity on the comb and positive electricity at the other end of this conductor, which is the comb opposite f. The effect of the points is to discharge these electricities upon the revolving plate, to be carried along by it, making the lower part negative and the upper part positive. The negative charge of the plate opposite f acts

Fig. 129.

inductively on f and induces a negative charge on it. The charges of the revolving plate act on the stationary plate through the air, and are in a measure "bound" except in those parts opposite the windows, where they are "free" and charge the armatures to higher potential. On separating m and n the conductor is broken, and if f and f' are not of sufficient difference of potential to act inductively on this parted conductor and effect a discharge across the break, the action of the machine will cease and the charges of f and f' will be dissipated through the air. To remedy this is the office of the conductor D. The positive charge of f', acting inductively on its lower end, draws negative electricity upon the plate, and in like manner f draws posi-

Fig. 130.

tive electricity upon the upper part of the plate. These charges go to build up f and f'. Hence, when the action through the horizontal conductors ceases, owing to the separation of m and n, the increase of the difference of potential between f and f' still goes on through the agency of D, to be aided, of course, by each discharge between m and n.

MAGNETISM AND ELECTRICITY. 195

300. **The Voss Machine** (Fig. 129) differs from the Holtz machine in that the windows are omitted, and metallic buttons are cemented to the front of the revolving plate, over which sweep small tinsel brushes, which are electrically connected with the armatures on the stationary plate. These brushes becoming electrified through friction, and being connected with the armatures, set up a difference of potential between them for some unknown reason, this difference being afterwards increased as explained in the Holtz machine.

301. **The Wimshurst Machine** (Fig. 130) consists of two varnished glass plates revolving in opposite directions. On the outside of these plates strips of tin-foil are cemented radially. Two conductors at right angles to each other extend obliquely across the plates, one at the front and the other at the back. These conductors terminate in brushes of tinsel, which, as the plates revolve, electrically excite the strips. The discharging part of the machine is in connection with two insulated conductors, provided with combs and connected to small Leyden jars. The distance between the balls can be regulated by means of insulated handles, connected to the discharging apparatus. This machine is superior to the Holtz and the Voss because it is less affected by moisture and does not reverse its polarity when in action.

X. EXPERIMENTS WITH ELECTRICAL MACHINES.

302. **Attraction and Repulsion.** — Exp. — Charge the electrophorus as directed. Before lifting the plate, place a handful of small bits of paper on it. Account for their flying off on raising the plate.

Exp. — Support a metallic plate on a block of wood; on it place the bladder-glass (134), and resting on this glass a second metallic plate. Connect the bottom plate to one conductor of the Holtz machine and the top one to the other. In the glass vessel put a handful of pith balls. Work the machine and account for the dancing of the pith balls.

Exp. — Fill a small glass funnel, having an aperture of about one-eighth of an inch, with fine dry sand. Support the apparatus in some suitable way. Notice that the sand runs out in a smooth fine stream. Pass one end of a wire into the sand in the funnel, and connect the other end with one of the conductors of an electrical machine. Set the machine in action, and observe how the sand of the escaping stream scatters. Explain.

303. Lichtenberg's Figures. — **Exp.** — Charge a Leyden jar positively, and with its knob trace a small circle on the electrophorus bed. Charge a second Leyden jar negatively, and with its knob trace a cross within the circle. Through a muslin bag shake a mixture of sulphur and red-lead from a height of several inches upon the figure. The red-lead will accumulate around the cross and the sulphur around the circle (Fig. 131). What must have been the electrical condition of the sulphur and of the red-lead? How did they become electrified?

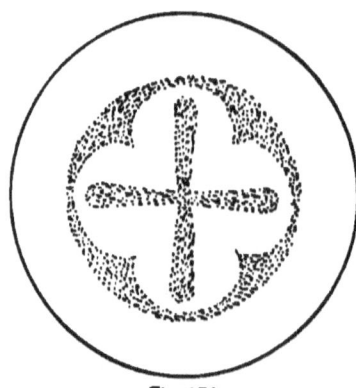

Fig. 131.

304. Action of Points. — **Exp.** — Suspend two pith-balls side by side from one of the conductors of an electrical machine. On working the machine, the balls separate widely. Why? Now hold the point of a needle near the conductor; the balls drop, showing that the conductor is discharged. Explain.

Exp. — Fasten a short, pointed wire to one of the conductors of an electrical machine, so that the point projects. Try to charge the machine. Account for the failure to obtain sparks between the conductors. Hold the flame of a candle near the point. It is driven away as by a wind. Account for this air-current.

Exp. — Connect an *Electric Whirl* (Fig. 132) to one of the conductors of an electrical machine. Notice the shape of its arms. Work the machine; the whirl rotates rapidly. Why?

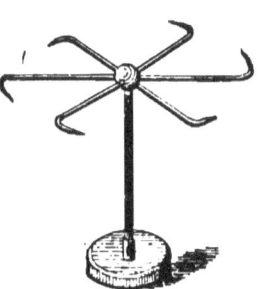

Fig. 132.

305. Mechanical Effects of Discharge. — **Exp.** — Between the discharging conductors of an active electrical machine, hold a piece of cardboard. It is perforated by the passing spark. A thin dry plate of glass may be perforated in like manner by a very heavy charge.

306. Heating Effects. — **Exp.** — Charge a Leyden jar. Connect its outer surface to a gas-burner by a chain or wire. Turn on the gas and bring the ball of the Leyden jar near enough to the mouth of the burner for a spark to pass. The gas will be lighted by the discharge. Explain.

Exp. — Fire ether in an iron spoon by proceeding as in the last experiment.

Exp. — Make a torpedo as follows: Fit corks to the ends of a paper tube, about 5 cm. long and 2 cm. diameter. Through one of them thrust two pieces of copper wire, the ends within the tube not quite touching. Fill the tube with fine gunpowder, close it, and place at a safe distance. Discharge a heavily-charged Leyden jar through it, using long wires for connections. Account for the result.

307. — Many interesting experiments illustrating other effects of the discharge or electric current are possible, such as those showing chemical, magnetic, physiological, or luminous effects. These will be given in a subsequent part of this chapter.

XI. ATMOSPHERIC ELECTRICITY.

308. Lightning. — It was demonstrated by Franklin in 1752 that lightning is identical with the electric spark. He sent up a kite during a passing storm, and found that as soon as the string became wet long sparks could be drawn from a key attached to the string, Leyden jars could be charged, and all the effects produced that characterize electricity. Lightning is but an electric discharge between two clouds, or between a cloud and the earth.

309. Kinds of Lightning. — Three kinds of discharge are recognized:

1. *Chain-Lightning*, when the path of the discharge is a zigzag, often consisting of several parts, and branching out in a very erratic fashion. The form is due, perhaps, in part, to the uneven conductivity of the air through which the discharge takes place, the line of least resistance having the irregular form marked out by the discharge.

2. *Sheet-Lightning*, when the reflection of the flash from the clouds is seen. It is often observed near the horizon during the hot weather of summer, and hence has been called *Heat-Lightning*.

3. *Globular-Lightning*, when balls of fire are seen to move slowly along and finally explode with great violence. This form is very rare, and no satisfactory explanation of it has ever been offered.

310. Thunder is the sound which follows a flash of lightning. The air is heated along the line of discharge producing sudden expansion and compression. This is followed by a violent rush of the air into the partial vacuum produced. When the path of the lightning is short and straight, there is a sharp "*clap*" like an explosion; when long and zigzag, there is the well-known "*rattle.*" The rumbling or rolling sound is due to echoes, the reflections of the sound by the clouds and other objects.

311. The Lightning-Rod is a metallic conductor placed on buildings to shield them from the effects of a discharge from the cloud. To be effective it should fulfil the following conditions:

1. The upper end should have several branches terminating in sharp points, and elevated above the highest part of the building.

2. The lower end should pass down into the earth till it meets with a good conducting stratum, and be deep enough to avoid damage to foundations and gas and water mains.

3. The conductor should be perfectly continuous, and of sufficient size to resist the heating effect of any discharge through it. Recent experiments show that iron is as suitable as copper for the purpose, and that the conductor should be either a flat strip or a bundle of wires with the several strands somewhat separated.

4. All metallic parts of the roof should be connected with the main conductor, and no insulators should be employed in securing the conductor in position. Maxwell suggested that a building should be covered with a network of wires, so that the building would form, as it were, the interior of a conductor (279).

312. Theory of the Rod. — If a highly electrified cloud is over any portion of the earth's surface, it charges all objects below it by induction with the opposite electricity. If these objects are armed with points, this electricity escapes from them, neutralizing that of the cloud. If the neutralizing action takes place too slowly to keep down the rapidly rising potential and prevent the disruptive discharge, the rod should offer the line of least resistance and protect the building. But it is now known that the discharge does not always follow the path of least resistance. In such cases the rod may not insure protection. Its protective agency is probable, however, if not absolute. In any case, the area protected by a single point is very limited; a building should therefore be provided with many points placed at angles and projecting corners.

313. The Aurora is a luminous phenomenon which occurs in the regions of the poles. That seen in northern latitudes is called the *Aurora Borealis*, or *Northern Lights;* that in southern, the *Aurora Australis*, or *Southern Lights.* The aurora is probably due to electric discharges through the higher and thinner portions of the atmosphere, the cold air of the poles differing in potential from the warmer and moister air coming up from the tropical regions.

XII. CURRENT ELECTRICITY.

314. The Electric Current. —When an electrified body is discharged through a conductor, there is produced in it an electrical state called the *Electric Current* (284). Electrification is probably a state of strain (272); the electric current rapidly transfers this strain through the discharging conductor. If this strained condition is reproduced by the generator as fast as it is relieved by the conductor, the result is a continuous current. The devices

MAGNETISM AND ELECTRICITY. 201

thus far considered are not well adapted to maintain the condition described by the expression "difference of potential;" moreover, the electrical capacity of bodies is so small that the currents obtained are insignificant compared with those produced by the voltaic battery.

315. The Simple Voltaic Cell. — Exp. — Cut a strip of heavy sheet zinc and one of sheet copper, each about 10 cm. long and 3 cm. wide. Scour the zinc with emery paper till it is bright. Support these strips side by side in a glass tumbler two-thirds full of dilute sulphuric acid (one part acid to twenty of water). On touching the strips together, a shower of bubbles of gas will rise from the copper strip, and some also from the zinc. This gas can be shown to be hydrogen. Remove either strip, or do not allow them to touch, and the chemical action is much diminished. If a little mercury be now rubbed on the zinc strip, no gas will be given off by it; but if the upper ends of the two strips be connected by any of the substances in the list of good conductors (271), gas will again come off freely from the copper. This action will cease if the connection be made by any of the list of non-conductors. If the action is continued for some time, the zinc will be found to waste away, while the copper is unaffected.

Such a combination of two metals, immersed in a liquid which acts chemically on one of them, when they are connected by a conductor, constitutes what is known as a *Voltaic Cell* or *Element*. The name is derived from an Italian physicist, Volta, who first described such a cell in 1800. On applying the proper tests it is found that an electric current flows through the conductor from the copper plate to the zinc during the continuance of the chemical action.

316. Chemical Action in the Voltaic Cell. — Each molecule of sulphuric acid consists of seven atoms, of which two are hydrogen (H_2), one sulphur (S), and four oxygen (O_4). When amalgamated zinc is placed in the dilute acid under the conditions in the last article, the following chemical action takes place:

$$Zn + H_2SO_4 = Zn\,SO_4 + H_2.$$

Zinc and sulphuric acid produce zinc sulphate and hydrogen.

This action takes place at the surface of the zinc, but the hydrogen appears at the surface of the copper. The hydrogen is thus transferred through the liquid, not as free hydrogen gas, but by a succession of intermolecular exchanges, taking place when the two plates are electrically connected and a current of electricity is flowing through the circuit, from the copper to the zinc outside of the cell, and from the zinc plate to the copper through the liquid of the cell.

The hydrogen is thus transferred through the cell in the direction in which the current is flowing, while the remainder (SO_4) of each sulphuric acid molecule is transferred the other way. The action may then be represented as follows:

$$Zn \quad \overbrace{H_2\,SO_4 \;\mid\; H_2\,SO_4} \quad \overbrace{H_2\,SO_4 \;\mid\; Cu}$$

$$\longrightarrow$$

The arrow shows the direction of the current through the liquid, and the braces show the chemical connections to be made by the first step in the transfer. After this first step the chain of molecules becomes

$$Zn\,SO_4 \;\mid\; H_2\,SO_4 \;\mid\; H_2\,SO_4 \;\mid\; H_2\text{-}Cu.$$

It will thus be seen that zinc sulphate has been formed at the zinc plate, sulphuric acid has disappeared, and hydrogen gas has been set free at the copper plate. This

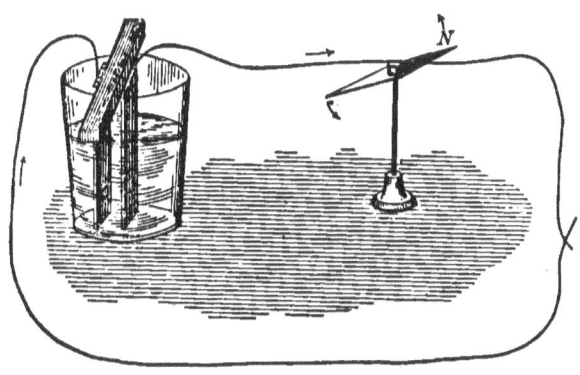

Fig. 133.

process is repeated indefinitely as long as the current continues to flow. It is known as the theory of Grotthus.

317. Oersted's Discovery. — **Exp.** — Solder a copper wire to each of the strips of the voltaic cell (315). Stretch a portion of the wire over a mounted magnetic needle (Fig. 133), holding it parallel to it and as near as possible without touching. Now bring the

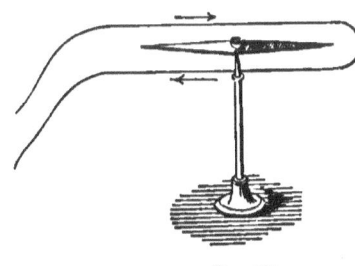

Fig. 134.

free ends of the wires together and observe that the needle is deflected, and after a few oscillations comes to rest at an angle with the wire. Next form a rectangular loop of the wire, and place the needle within it (Fig. 134). A greater deflection is now obtained. If a loop of several turns is formed, the deflection is still greater.

This experiment, first performed by Oersted in 1819, shows that the region round the wire has magnetic properties during the flow of electricity through it. The magnetic needle employed in this way becomes a *Galvanoscope*, a detector of electric currents.

318. Current-Direction Detected. — Exp. — Using the apparatus of Art. 317, compare the direction of the current through the wire with that in which the N-seeking pole of the needle turns. Cause the current to pass in the reverse direction over the needle; the deflection is reversed. Now hold the wire below the needle, and the direction of deflection is again reversed.

The direction of deflection may always be predicted by the following rule: *Place the palm of the right hand next the wire, but on the side opposite the needle, so that the outstretched fingers point in the direction of the current. Then the outstretched thumb will point in the direction of deflection of the N-seeking pole of the needle.*

319. The Circuit of a voltaic cell comprises the entire path traversed by the electric current, including the conducting plates and liquid within the cell, as well as the external conductor. The plate which wastes away by chemical action, usually zinc, is called the *negative* plate, or *electrode*, and the other is the *positive*. *Closing the circuit* means joining the two electrodes by a conductor; *breaking the circuit* is disconnecting them. The current flows in the external circuit from the positive electrode to the negative.

320. Local Action. — Exp. — Place a strip of commercial zinc in dilute sulphuric acid. Hydrogen is liberated during the chemical action, and after a few minutes the zinc becomes black from particles of carbon exposed to view on dissolving away the surface. If the experiment is repeated with both *pure* zinc and zinc amalgamated with mercury (315), there will be little or no chemical action.

The experiment shows that the amalgamation of commercial zinc with mercury imparts to it properties similar to those of pure zinc. If in the experiment with the simple voltaic cell, a galvanoscope is inserted in the circuit both before the zinc has been amalgamated and afterwards, it will be found that a larger deflection will be obtained in the second case.

The chemical action going on in a voltaic cell which contributes nothing to the current flowing through the circuit is known as *Local Action*. It is probably due to the presence of particles of carbon, iron, etc., in the zinc; and these with the zinc form small voltaic cells, the currents flowing round in short circuits from the zinc through the liquid to the foreign particles and back to the zinc again. Adjacent hard and soft portions of the zinc act in a similar way.

This local action is prevented by amalgamating the zinc; that is, by coating it with an alloy of mercury and zinc. The amalgam brings pure zinc to the surface, covers the foreign particles, and above all forms a smooth surface, so that a film of hydrogen clings to it and protects it from chemical action save when the circuit is closed.

321. Polarization. — **Exp.** — Connect the poles of the voltaic cell to the galvanoscope and note the amount of deflection. Let the cell remain in circuit with the galvanoscope for some time, the deflection will gradually become less and less. Now stir up the liquid vigorously with a glass rod, inserting it between the plates and brushing off the adhering gas bubbles; the deflection will increase to nearly its original amount.

This diminution in the intensity of the current is due to several causes, but the chief one is the film of hydrogen which gathers on the copper plate, causing what is known as the *Polarization* of the cell. The hydrogen on the positive plate not only introduces more resistance to obstruct the flow of the current, but it also diminishes the electromotive force (352) to which this flow is ascribed. The origin of the electromotive force ($E.M.F.$) is the superior affinity of zinc for oxygen over that of copper. Since hydrogen has an affinity for oxygen greater than that possessed by copper, its presence on the copper plate sets up an inverse $E. M. F.$ which either reduces or stops the flow of the current.

Exp. — Place enough mercury in a quart jar to cover the bottom, and hang above it a piece of sheet zinc. Fill the jar with a nearly saturated solution of salt water, and place in the mercury the exposed end of a copper wire insulated with gutta percha, the upper end forming the positive pole of the battery.

If now the circuit is closed through a telegraph sounder (386) or relay (387) of about fifty ohms resistance, the armature will at first be attracted strongly; but in the course of a few minutes it will be released and will be drawn back by the spring. Polarization has then set in to the extent that the current is insufficient to operate the instrument.

Next take a small piece of mercuric chloride ($Hg\ Cl_2$) no larger than the head of a pin, and drop it in on the surface of the mercury. The armature of the sounder will instantly be drawn down, showing that the current has recovered its normal value. The hydrogen has been removed by the chlorine of the mercuric chloride. In a few minutes the chlorine will be exhausted, and polarization will again set in. A little more of the chloride will again restore the activity of the cell.

322. **The Remedies for Polarization** are two in number: 1. Roughening the positive plate to lessen the adhesion of the gas. 2. The use of some oxidizing agent, as nitric acid, chromic acid, or manganese dioxide, to oxidize the hydrogen before it reaches the positive plate. Attempts to prevent polarization have given rise to a great many forms of cells or batteries.

323. The Smee Cell consists of a plate of silver or lead between two zinc plates, dipping into a glass vessel containing dilute sulphuric acid (Fig. 135). Polarization is in a measure prevented by roughening the positive plate by means of a coating of finely-divided platinum. This coating also raises the electromotive force. The remedy, however, is but partial.

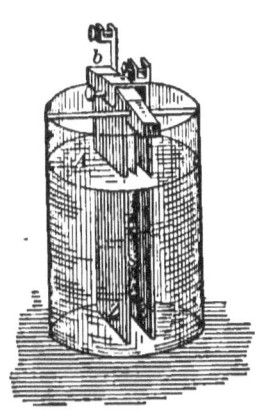

Fig. 135.

324. The Daniell Cell (Fig. 136), in its most common form, consists of a glass vessel containing a solution of copper sulphate in which stands a sheet of copper bent into a cylindrical form. Within the copper cylinder is a porous vessel of unglazed earthenware containing dilute sulphuric acid, or, preferably, a solution of zinc sulphate. In this porous cup stands a zinc prism. To the top of the copper plate is often fastened a perforated pocket in which are placed crystals of copper sulphate to keep the copper sulphate solution saturated.

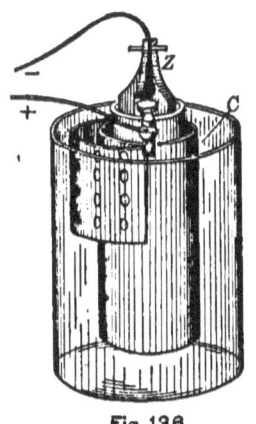

Fig. 136.

When the circuit is closed, the action of the acid on the zinc liberates hydrogen which travels toward the porous cup by a series of molecular exchanges, as explained in Art. 316. At the porous cup it meets the copper sulphate and supplants the copper atoms, forming sulphuric acid. The copper then travels in like manner with the current, and is finally deposited on the copper plate. The action may be represented thus:

$$Zn_x \mid H_2SO_4 \mid H_2SO_4 \parallel CuSO_4 \mid CuSO_4 \mid Cu_y$$

⟶

After the first step in the reaction this becomes

$$Zn_{x-1} \mid ZnSO_4 \mid H_2SO_4 \parallel H_2SO_4 \mid CuSO_4 \mid Cu_{y+1}$$

The arrow shows the direction of the current, and the double vertical lines represent the porous cup. It will be seen that the zinc has lost one atom, and the copper has gained one. Also one molecule of zinc sulphate has been formed, and one molecule of copper sulphate has disappeared. The hydrogen is intercepted by the $CuSO_4$, and never reaches the copper plate. Polarization is thus entirely prevented, and the cell is one of the most constant known.

Fig. 137.

325. **The Gravity Cell** (Fig. 137) is a modified Daniell. The porous cup is omitted, the partial separation of the liquids being secured by difference in density. The copper plate is placed at the bottom in saturated copper sulphate, while the zinc is suspended near the top in a weak solution of zinc sulphate, floating on top of the copper sulphate. The zinc should never be placed in the solution of copper sulphate. The saturated copper sulphate is more dense than

the dilute zinc salt, and so remains at the bottom, except as it slowly diffuses upwards.

326. **The Grove Cell** (Fig. 138) consists of a glass vessel containing dilute sulphuric acid in which is immersed a hollow cylinder of zinc. Within this zinc cylinder is a porous cup containing strong nitric acid and a strip of platinum.

Fig. 138.

The hydrogen produced by the action of the acid on the zinc meets with the nitric acid at the porous cup, and is oxidized to water. The nitric acid molecule is broken up by this action, yielding a brownish-red gas, which is very corrosive and poisonous.

327. **The Bunsen Cell** (Fig. 139) differs from the Grove cell in the substitution of carbon for the platinum. The chemical action is the same. To avoid the poisonous fumes a solution of chromic acid in dilute sulphuric acid is sometimes substituted for the nitric acid. Then the chemical action is similar to the chromic acid battery.

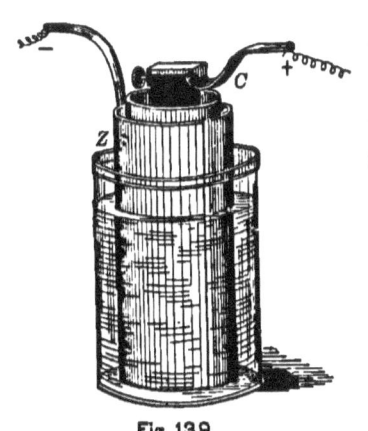

Fig. 139.

328. **The Chromic Acid Cell** usually consists of a plate of zinc between two carbon plates dipping into a glass vessel containing dilute sulphuric acid to which is added

either chromic acid or the bichromate of potassium or of sodium. The sodium salt is much to be preferred to the potassium. With the bichromates an additional quantity of acid is needed to liberate chromic acid.

Fig. 140 illustrates a form of this battery called the *Grenet* or *Bottle Battery*, which is very convenient, but is open to the objection that since the carbon plates are left standing in the solution, the liquid soon works up and attacks the connections at the top, making it difficult to keep the cell in good order. The zinc is attached to a sliding rod, so that it may be lifted out of the liquid when the battery is not in use.

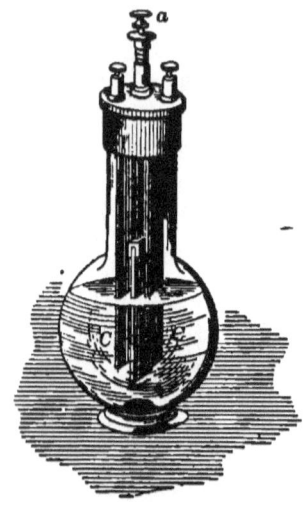

Fig. 140.

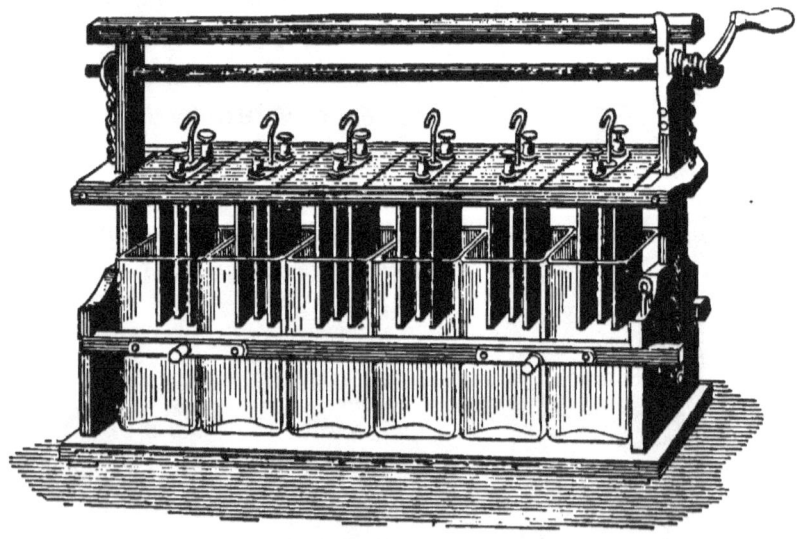

Fig. 141.

MAGNETISM AND ELECTRICITY. 211

The hydrogen evolved by the action of the acid on the zinc is oxidized to water by the chromic acid, and polarization is prevented.

Fig. 141 illustrates a form of chromic acid battery, where the several cells composing it have their carbons and zincs suspended from a frame. It is known as a *Plunge Battery*, and is a very convenient form for experimental work.

329. The Leclanché Cell (Fig. 142) consists of a glass vessel containing a saturated solution of ammonium chloride (sal ammoniac) in which stands a zinc rod and a porous cup. In this porous cup is a bar of carbon very tightly packed in a mixture of manganese dioxide and graphite, or granulated carbon.

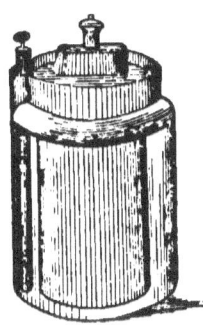

Fig. 142.

The zinc is acted on by the chlorine of the ammonium chloride, liberating ammonia and hydrogen. The ammonia in part dissolves in the liquid, and in part escapes into the air. The hydrogen is slowly oxidized by the manganese dioxide. The cell is not adapted to continuous use, as the hydrogen is liberated faster than the oxidation goes on, thereby polarizing the battery. If, however, it is allowed to rest, it recovers from polarization.

XIII. EFFECTS OF ELECTRIC CURRENTS.

a. HEATING EFFECTS.

330. Heating Effects. — **Exp.** — Close the circuit of a chromic acid battery through a piece of No. 30 platinum wire about 3 cm. long. The wire becomes red hot and possibly may fuse. If copper wire is substituted for the platinum, a smaller change of temperature will be observed.

In a battery, the potential energy of chemical separation is transformed into the energy of an electric current. When the current does no work this energy is all converted into heat in the circuit. The relative amounts of heat generated in the external circuit and internally in the battery itself are proportional to the external and internal resistances. If in the experiment the circuit is closed with the fine wire omitted, more heat is generated in the liquid of the battery than with the fine wire in the circuit (347).

331. Laws of the Development of Heat by a Current. — Joule demonstrated the truth of the following laws:

I. *The heat developed is proportional to the square of the current strength.*

II. *In any portion of a circuit it is proportional to the resistance of that portion.*

III. *The heat is also proportional to the time during which the current flows.*

The thermal effects of the electric current are utilized in firing blasts and cannons, in exploding torpedoes, in cauterizing, in electric heating, and in electric lighting.

b. CHEMICAL EFFECTS.

332. Electrolysis. — **Exp.** — Bend a glass tube of about 1.5 cm. diameter and 15 cm. long into a V-form (Fig. 145). Close the ends with corks and thrust through them platinum wires, terminating within the tube in narrow strips of platinum foil. Support the tube in some convenient way after filling it two-thirds full of a solution of sodium sulphate, colored with the extract of purple cabbage. Connect the terminals to the poles of two or three cells joined in series (355). On closing the circuit for a few minutes, the liquid around the positive pole, where the current enters, turns red, showing the presence of an acid, while that around the negative pole turns green, showing the presence of an alkali.

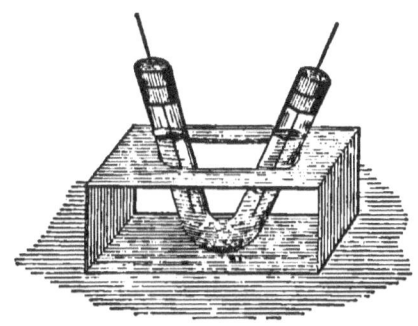

Fig. 143.

The experiment shows that the passage of an electric current through a compound liquid decomposes it. To this process of decomposing liquids by means of an electric current Faraday gave the name of *Electrolysis;* to the substance decomposed, *Electrolyte;* to the parts of the separated electrolyte, *Ions*.

333. Electrolysis of Copper Sulphate. — Exp. — Fill the V-tube of the last experiment about two-thirds full of a solution of copper sulphate. After the circuit has been closed a few minutes, the negative electrode will be covered with a deposit of copper and bubbles of gas will rise from the positive electrode. These bubbles are oxygen.

The action has been to separate the copper sulphate into copper and sulphion (SO_4). By the process of molecular exchanges already explained in connection with batteries (316), the copper is transferred along with the current and the sulphion against the current. At the positive pole, where the sulphion would be liberated, it abstracts hydrogen from water, forming sulphuric acid, and sets free oxygen, which then comes off as bubbles of gas. If the positive electrode were copper the sulphion would unite with it, reforming copper sulphate at this pole as fast as it is decomposed at the other.

334. Electrolysis of Water. — Exp. — Cut the bottom from a wide-mouthed bottle. Insert a good cork in the mouth, and through it thrust two platinum wires terminating within the bottle in strips of platinum foil set parallel to each other, and about two centimetres apart (Fig. 144). Fill the bottle two-thirds full of water acidulated with sulphuric acid, and support it in the ring of the iron stand.

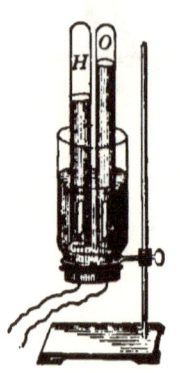

Fig. 144.

Over these electrodes invert test-tubes filled with acidulated water. Place the apparatus in circuit with two or more chromic acid cells connected in series (355). Bubbles of gas at once begin to rise from the electrodes. If the action be allowed to continue for a few minutes the tube over the negative electrode will be seen to contain about twice as much gas as the other. This can be shown to be hydrogen, the other oxygen.

During the passage of the current the hydrogen of the sulphuric acid moves with the current, and the sulphion (SO_4) against it, and finally releases oxygen from the water by taking away the hydrogen to re-form H_2SO_4. If copper or brass electrodes be used, the sulphion will attack the positive one and no oxygen will appear.

335. **Electroplating** consists in covering bodies with coatings of any metal by means of the electric current. The process may be summarized as follows. Thoroughly clean the surface to remove all fatty matter. Attach the article to the negative electrode of a battery, the electrolyte being a solution of some chemical salt of the metal to be deposited. If silver, cyanide of silver is used; if copper, sulphate of copper. To maintain the strength of the solution a piece of the metal of the kind to be deposited is attached to the positive electrode. The action is similar to that heretofore given. Articles of iron, steel, zinc, tin, and lead cannot be silvered or gilded unless first covered with a thin coating of copper.

336. Electrotyping consists in copying medals, woodcuts, type, and the like in metal, usually copper, by means of the electric current. A mould of the object is taken in wax or plaster of Paris. This is evenly covered with powdered graphite to make the surface a conductor, and treated very much as an object to be plated. When the deposit has become sufficiently thick it is removed from the mould and backed or filled in with type-metal.

337. The Secondary or Storage Battery.— Exp. — Connect the apparatus of Art. 334 to a suitable battery. After passing the current for a short time, causing an evolution of gas, disconnect the battery and put a galvanoscope in its place. The needle will be deflected, showing that a current is now passing through the apparatus in a direction opposite to the one which produced the electrolysis.

If large plates of lead are used as electrodes, and dilute sulphuric acid as the electrolyte, we have an arrangement which illustrates the construction of the *Storage Battery of Planté*. The electrolysis of the water liberates oxygen on one plate, which combines with the lead to form a deposit of lead peroxide (PbO_2). Hydrogen accumulates on the other plate. On disconnecting the battery and connecting the lead plates by a conductor, a current flows from the oxidized plate to the other one, the lead peroxide is reduced to spongy lead on the positive plate, while more or less lead sulphate is formed on the negative.

The storage battery stores energy instead of electricity. The energy of the charging current is converted into the potential energy of chemical separation in the storage cell. When the circuit of the secondary cell is closed the potential chemical energy is reconverted into the energy of an electric current in precisely the same way as with a primary battery.

c. MAGNETIC EFFECTS.

338. Magnetic Character of the Current. — **Exp.** — Connect two or three chromic acid cells in parallel circuit (356).

Fig. 145.

Close the circuit through a heavy wire, and then dip a portion of it into fine iron filings. A thick cluster of them will adhere to the wire (Fig. 145).

The magnetic properties of a conductor carrying an electric current were also seen in its action on the magnetic needle (317).

339. Magnetic Field about Conductor. — **Exp.** — Support horizontally a sheet of stiff paper. Pass vertically through it a wire which connects the poles of a battery of two or more chromic acid cells (Fig. 146). Close the circuit and sift a few very fine iron filings on the paper, jarring it slightly with a pencil as they fall. They arrange themselves in circles with the wire at the centre. Place a small mounted magnetic needle on the paper near the wire. The needle sets itself tangent to these circles, and points in the opposite direction to that traversed by the hands of a watch when the current enters the plane of the paper from below. (What is the direction of the lines of force?)

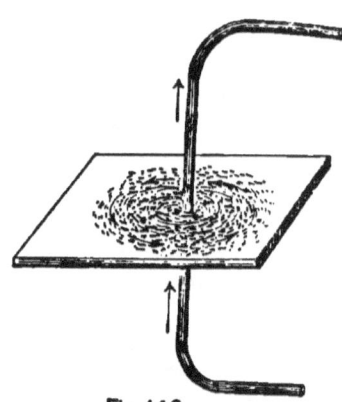

Fig. 146.

The experiment shows that the lines of force of this magnetic field are concentric circles. Their direction is given by the following rule:

Hold the closed right hand so that the extended thumb

points in the direction of the current; then the fingers will indicate the direction of the lines of force.

340. Properties of a Circular Conductor. — Exp. — Bend a piece of about No. 16 copper wire into the form shown in Fig. 147, the diameter of the circle being about 20 cm. Suspend it by a long thread, so that the ends dip into the mercury cups shown in section in the lower part of the figure. Send a current through the suspended wire by connecting a battery to the binding posts. A magnet brought near the face of the circular conductor will cause the latter to turn about a vertical axis and take up a position at right angles to the axis of the magnet.

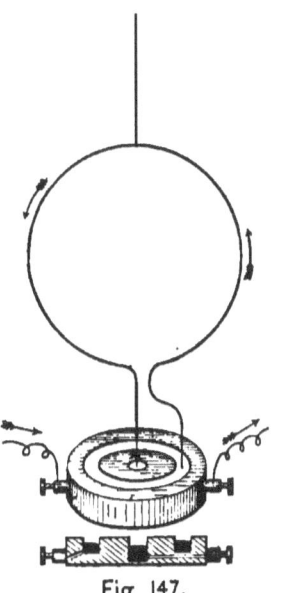

Fig. 147.

This experiment, due to Arago, shows that a circular current acts like a disk magnet, whose poles are its faces. The lines of force surrounding the conductor in this form pass through the circle and come round from one face to the other through the air outside the loop. The N-seeking side is the one from which the lines issue; and to an observer looking toward this side, the current flows round the loop counter clockwise (Fig. 148).

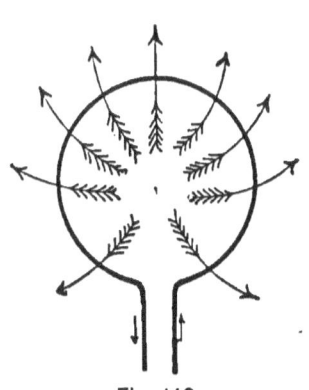

Fig. 148.

341. The Helix. — If instead of a single turn of wire we take a long insulated wire and coil it into a circle of several layers, the magnetic effect will be increased. Such a coil is called a *Helix* or *Solenoid;* and the passage of an electric current through it gives to it all the properties of a cylindrical bar magnet.

342. The Electro-Magnet. — Exp. — Wind neatly on a paper tube, of about 2 cm. diameter and 15 cm. long, three layers of No. 18 insulated copper wire. Pass an electric current through it and test its magnetic properties by bringing it near a mounted magnetic needle. Now fill the tube with straight pieces of soft iron wire and again bring it near the needle. Its magnetic effect will be greatly increased.

This device, consisting of a helix encircling an iron core,

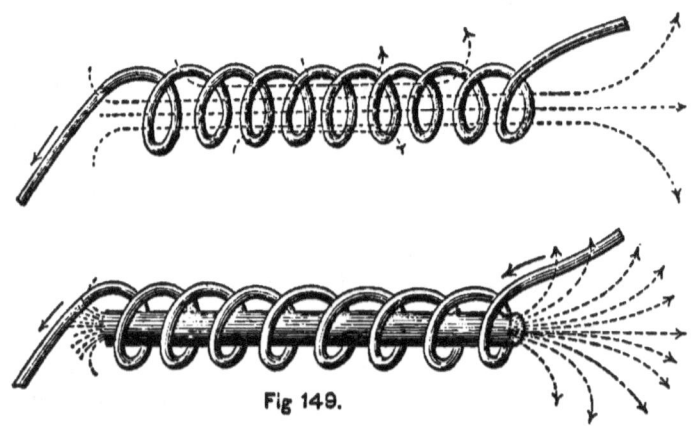

Fig 149.

is called an *Electro-Magnet*. The presence of the iron core greatly increases the number of lines of force running through the helix from end to end, by reason of its permeability (257) or its conductivity for lines of force (Fig. 149). When a core is not used, many of the lines leak out at the sides of the helix, and but few extend from end to end. The core not only diminishes the leakage of the lines of force, but also adds many more to those previously existing. Hence the magnetic strength of a helix is greatly increased by the iron core.

MAGNETISM AND ELECTRICITY. 219

343. **The Form** given to an electro-magnet depends on the use to which it is to be put. The *Horseshoe* or *U-shaped* (Fig. 150) is the most common. The advantage of this form becomes apparent when attention is directed to the fact that *every magnetic circuit is a closed one;* that is, the lines of magnetic induction are continuous, passing through the core from the south to the north pole, and

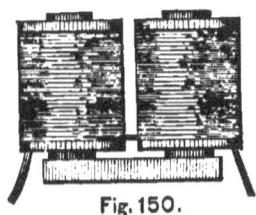

Fig. 150.

completing the circuit by passing through the air from the north pole back to the south. Now the shorter the air path of the magnetic lines, the larger their number and the stronger the magnet. The approach of the two poles in the U-shaped magnet shortens the air gap and hence increases the number of lines of force. If now a bar of soft iron, called an *Armature*, is placed across the poles, the air gap is further diminished when the armature is made to approach the poles (Fig. 151). With the armature in contact with them the magnetic circuit is all iron, and the maximum number of magnetic lines traverse it (Fig. 152).

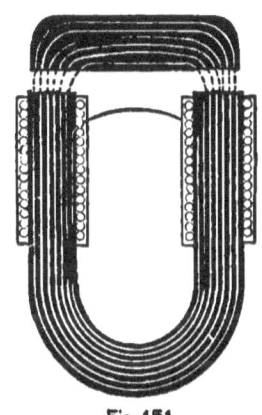

Fig. 151.

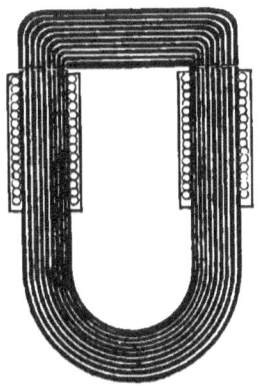

Fig. 152.

344. **The Polarity of a Solenoid** may be determined by the following rule:

Grasp the coil with the right hand

so that *the fingers point in the direction of the current;* the north pole will then be in the direction of the extended thumb.

In Fig. 153, if the fingers of the right hand are parallel to the arrows and point in the same direction, the extended thumb will point toward N, as the north pole of the coil. The converse of the rule enables one to ascertain the direction of the current, when the polarity of the solenoid is known.

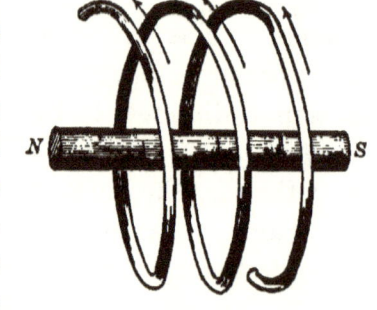

Fig. 153.

345. Mutual Action of Two Currents. — Exp. — Construct a rectangular frame, about 25 cm. square, out of insulated copper wire No. 20, by winding four or five layers around the edge of a square board. Slip the wire off the board and tie the parts together in a number of places with thread.

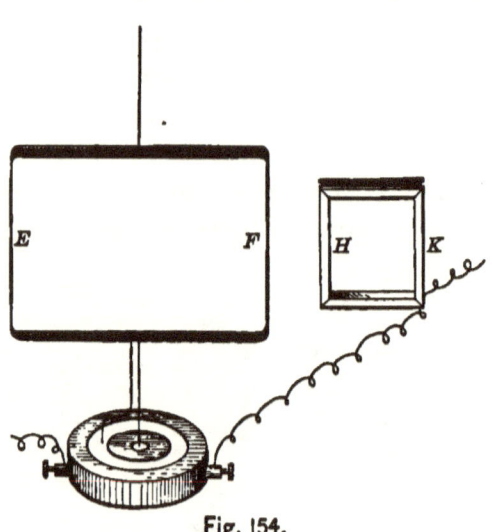

Fig. 154.

Bend the ends at right angles to the frame, remove the insulation from them, and give them the form shown in Fig. 154. Suspend the wire frame by a long thread, so that the ends dip into mercury cups like those shown in section in Fig. 147.

Wind several layers of the same kind of insulated wire around a frame or the edge of a board about 10 cm. by 20 cm. Connect this coil in the same circuit with the rectangular coil and a battery of two or three cells joined in series (355).

First. Hold the coil HK with its plane perpendicular to the plane of the coil EF, with its edge H parallel to F, and with the currents in these two adjacent portions flowing in the same direction. The suspended coil will turn upon its axis, the edge F approaching H, showing attraction.

Second. Turn HK over so that the currents in the adjacent portions H and F flow in opposite directions. The edge F of the suspended coil will be repelled by H.

Third. Give HK a quarter turn around a vertical axis from the position shown in the figure. If it is near F, the suspended coil will turn in a direction to make the two currents parallel and in the same direction round.

Fourth. Hold the coil HK within the rectangular coil EF, and with the edge H making an angle with the lower edge of EF. The coil EF will turn till the currents in its lower edge are parallel with those in H, and flowing in the same direction.

These results may be expressed by the following *Laws of Action between Currents*:

I. *Parallel currents flowing in the same direction attract.*

II. *Parallel currents flowing in opposite directions repel.*

III. *Currents making an angle with each other tend to become parallel and to flow in the same direction.*

Maxwell included the entire phenomenon under one law, viz.: *The two circuits tend to move so that they shall include the largest number of lines of force common to the two.*

XIV. ELECTRICAL QUANTITIES.

a. OHM'S LAW.

346. Frequent mention has been made in the preceding pages of certain electrical magnitudes, such as *Resistance*, *Electromotive Force*, and *Strength of Current*. We shall now define them more fully and explain the relation existing between them, which is known as *Ohm's Law*. This law forms the basis of most electrical measurements of steady currents.

347. Resistance. — No conducting body possesses perfect conductivity, but every conductor presents some obstruction to the passage of electricity. This obstruction is called its *Electrical Resistance*. It is the reciprocal of *Conductivity*. The greater the conductivity of a conductor the less its resistance, the one decreasing in the same ratio as the other increases.

348. The Unit of Resistance. — The practical unit of resistance is the *Ohm*. It was defined by the International Congress of Electricians in Chicago, August 21, 1893, as follows: The international ohm is represented by the resistance offered to an unvarying electric current by a column of mercury at the temperature of melting ice, 14.4521 grammes in mass, of a constant cross-sectional area, and of the length 106.3 centimetres.

349. Laws of Resistance. — I. *The resistance of a conductor is proportional to its length.* For example, if 39 ft. of No. 24 copper wire (*B* and *S* gauge) have a resistance of 1 ohm, then 78 ft. of the same wire will have a resistance of 2 ohms.

II. *The resistance of a conductor is inversely proportional to its cross-sectional area, and in the case of round wire is inversely proportional to the square of its diameter.* For example, No. 24 wire has a diameter of .0201 inch and No. 30 has a diameter of .01 inch, or nearly one-half of No. 24. 39 ft. of No. 24 has a resistance of 1 ohm, and 9.7 ft. of No. 30, which is nearly $\frac{39}{2^2}$, has a resistance also of 1 ohm.

III. *The resistance of a conductor of a given length and cross-sectional area depends upon the material of which it is made, and is affected by any cause which modifies its molecular condition.* For example, the resistance of 2.68 ft. of German-silver wire No. 24 is 1 ohm, whereas it takes 39 ft. of copper wire of the same diameter to give the same resistance. Moreover, this is true only at a definite temperature; if these substances are heated, the resistance of the copper increases nearly ten times as much as that of German silver, and twenty times as much as an alloy called platinoid. All metals have their resistance increased by increase of temperature. Carbon and all electrolytic conductors decrease in resistance as the temperature rises.

Exp. — Connect the poles of a fresh chromic acid cell with a piece of fine platinum wire of such a length that the current heats it to a dull red color. Now apply a piece of ice to one portion and notice the rise in temperature of the remaining portion. Explain.

350. Formula for Resistance. — The above laws are conveniently expressed in the following formula:

$$r = k \cdot \frac{l}{d^2},$$

in which l represents the length of the wire in feet, d its diameter in thousandths of an inch, and k is a constant depending on the material and is of such a value that r is obtained in ohms. The following table gives the approximate value of k for several metals, when the temperature is 20° C.:

Silver	9.84	Zinc	36.69	German-silver	128.29
Copper	10.45	Platinum	59.02	Mercury	586.24
		Iron	63.35		

351. The Strength of a Current determines the magnitude of the effects produced by it. Either the chemical, electro-magnetic, or heating effects may be made the basis of a system of measurement. The quantity of an ion decomposed in a second furnishes a convenient magnitude for the determination of unit strength of current. The unit of current strength is the *Ampere*. It is that current which will deposit by electrolysis, under suitable conditions, 0.001118 gramme (0.017253 grain) of silver, or 0.0003287 gramme (0.005072 grain) of copper in one second. The ampere deposits 4.025 grammes of silver in one hour. A milliampere is a thousandth of an ampere. It is to be remarked that the electrolytic method measures only the quantity of electricity passing through the decomposing cell, called a *Voltameter*, in the given time.

352. Electromotive Force is the name given to the cause of an electric flow. It is often called *Electric Pressure* from its superficial analogy to water pressure. The unit of electromotive force (*E.M.F.*) is the *Volt*. It is the *E.M.F.* which will cause a current of one ampere to flow through a resistance of one ohm. The *E.M.F.* of a battery depends upon the materials employed, and is entirely independent of the size and shape of the plates. The *E.M.F.* of a Daniell cell is about 1.1 volts; of a fresh chromic acid cell, 2 volts; and of a Lelanché cell, 1.5 volts. The *E.M.F.* of a Carhart-Clark standard cell is 1.44 volts at 15° C.

353. Ohm's Law.— The definite relation existing between strength of current, resistance, and *E.M.F.* was first announced by Ohm in 1827. It is known as *Ohm's Law*, and may be expressed as follows:

The strength of a current equals the E.M.F. divided by the resistance; or in symbols:

$$C = \frac{E}{R},$$

where C is the current in amperes, E the E.M.F. in volts, and R the resistance in ohms. Applied to a battery, if R is the resistance external to the cell, and r the internal resistance of the cell itself, then

$$C = \frac{E}{r + R}.$$

Thus, if a chromic acid cell of 2 volts E.M.F. and half an ohm internal resistance is closed with a wire having a resistance of one and a half ohms the current will be $\frac{2}{1.5 + 0.5} = 1$ ampere.

From the equation $C = \frac{E}{R}$, we derive $E = CR$; that is, the effective E.M.F. equals the product of the current and resistance.

354. Methods of Varying Strength of Current. — It is evident from Ohm's law that the strength of the current furnished by an electric generator may be increased in two ways: 1. By lessening the resistance. 2. By increasing the E.M.F.

The E.M.F. may be increased by joining several cells in series, and the internal resistance may be diminished by connecting them in parallel. Enlarging the plates of a battery or bringing them closer together diminishes the internal resistance.

355. Connecting in Series. — If several cells are connected so that the positive pole of one is joined to the negative pole of the next and so on, then the total *E.M.F.* is the sum of the *E.M.F.'s* of the several cells. The cells are then said to be joined in *Series*. Fig. 155 represents the conventional sign for a single cell. The short, thick line represents the negative plate, and hence the line from it is the negative pole; the long, thin line represents the positive plate, and hence the line from it is the positive pole.

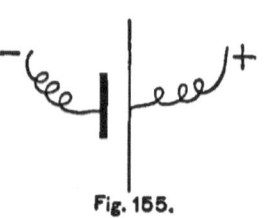

Fig. 155.

Fig. 156 represents six cells joined in series. If each cell has an *E.M.F.* of 2 volts, then the total *E.M.F.* will be 6×2 or 12 volts. In connecting cells in this manner the internal resistance is also increased about six times, since the liquid conductor is six times as long as in one cell. The current for any external resistance R is then $C = \dfrac{12}{6r + R}$.

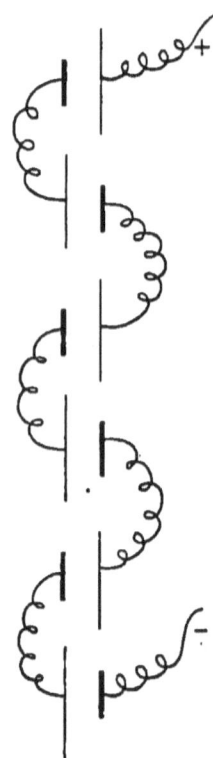

Fig. 156.

356. Connecting in Parallel.—When all the positive terminals are connected together on one side and the negative on the other, the cells are grouped in *Parallel* or *Multiple Arc*. With n similar cells the effect of such a grouping (Fig. 157) is to reduce the internal resistance to $\dfrac{1}{n}$th that of a single cell. It is equivalent to increasing the area of the plates n times. All the cells side by side contribute equal shares to the output of the battery.

357. Relative Advantages of the Series and Multiple Grouping. — It is evident from Ohm's law that when the external resistance is small, there is nothing gained by increasing E and at the same time increasing r, since then C remains practically unchanged. But when R is large, then the increase in r due to joining cells in series is more than counterbalanced by the increase in E. Consequently C is greater the larger the number of cells joined in series. For example: A battery of 6 cells, each having an *E.M.F.* of 2 volts and an internal resistance of 0.5 ohm, acts, first, through an external resistance of 0.1 ohm, and, secondly, through one of 500 ohms. If joined in parallel circuit, then when $R = 0.1$,

$$C = \frac{2}{0.1 + \frac{0.5}{6}} = 10.9 \text{ amperes, and}$$

when $R = 500$, $C = \dfrac{2}{500 + \frac{0.5}{6}}$

$= 0.004$ ampere. If joined in series, then when $R = 0.1$, $C = \dfrac{6 \times 2}{0.1 + 6 \times 0.5} = 3.87$ amperes, and when $R = 500$, $C = \dfrac{6 \times 2}{500 + 6 \times 0.5} = 0.024$ ampere. A comparison of these results shows that when the external resistance is small, the greater current is obtained by grouping in parallel; but when the external resistance is large, the series arrangement gives the greater current.

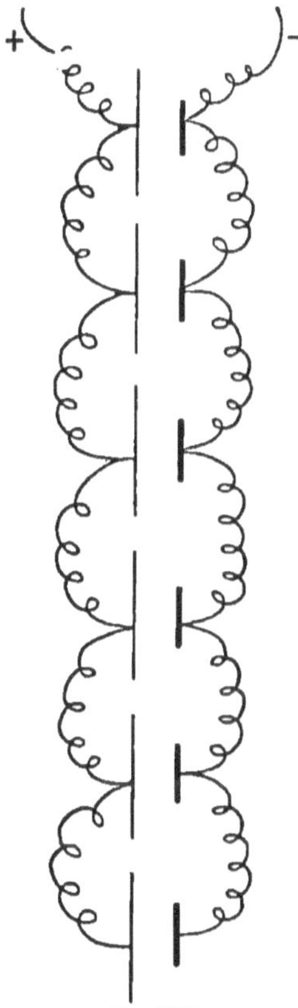

Fig. 157.

b. INSTRUMENTS FOR MEASUREMENT.

358. The Galvanometer. — If comparison of currents is made by means of their magnetic effects, the instrument used for the purpose is called a *Galvanometer*. If the galvanometer is calibrated, so as to read directly in amperes, it is called an *Ammeter*. A galvanoscope becomes a galvanometer by providing it with a scale so that the deflections may be measured. In very sensitive instruments a small mirror is attached to the movable part of the instrument; it is then called a *Mirror Galvanometer*. Sometimes a beam of light from a lamp is reflected from this small mirror back to a scale, and sometimes the light from a scale is reflected back to a small telescope, by means of which the deflections are read. In either case the beam of light then becomes a long pointer without weight.

359. The Tangent Galvanometer consists of a vertical coil of wire (Fig. 158) of from ten to twelve inches in diameter, at the centre of which is supported a magnetized needle about three-quarters of an inch long, furnished with a long, light pointer moving over a graduated scale. Owing to the size of the coil the magnetic field at the centre is nearly uniform, that is, the lines of force there are sensibly parallel straight lines; and any movement of the short needle round a vertical axis will not carry its poles into a magnetic field of different strength. Under these conditions *the strength of the current is pro-*

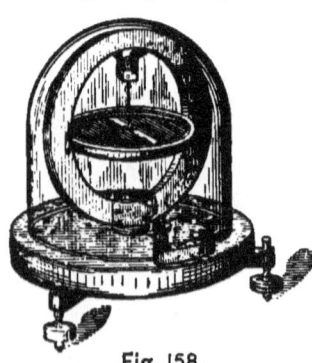

Fig. 158.

portional to the tangent of the angle of deflection. For example, if two different batteries, placed successively in circuit with a tangent galvanometer, give deflections of 55° and 35½° respectively, then the strengths of these currents are as $\dfrac{tan\ 55°}{tan\ 35\frac{1}{2}°} = \dfrac{1.428}{0.714} = 2$, that is, the strength of one current is double that of the other.

In its most useful form the helix consists of several coils, so connected to suitable binding-posts that any or all can be placed in circuit, thus making the instrument either a *long* or a *short coil* galvanometer at pleasure.

360. The d'Arsonval Galvanometer. — One of the most useful forms of galvanometer is the d'Arsonval, Fig. 159. Between the poles of a strong permanent magnet swings a coil suspended by a fine wire in such a way that the current is led in by the suspending wire and out by the wire connecting the coil to a spring at the bottom. Inside the coil is a soft iron tube supported from the back. In some of the most recent forms of this instrument the coil is made narrower and the iron tube is omitted. In

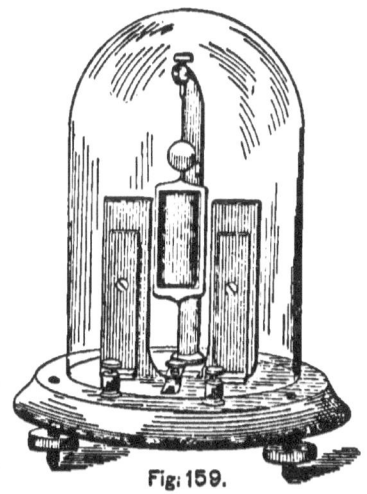

Fig. 159.

this galvanometer the coil is movable and the magnet is fixed. Its chief advantages are simplicity of construction, almost total independence of the direction and strength of the magnetic field at the place where it is used, and the quickness with which it comes to rest after the coil has been deflected by a current through it.

361. Resistance Coils.—Coils of wire of known resistance, for use with the galvanometer in measuring resistance, can be purchased of makers of electrical instruments. They should be mounted as shown in Figure 160. The coil is wound on the spool double, the inner ends being connected so that the current passes an equal number of times in both directions round the spool, and the coils do not become magnets. This winding also diminishes what is known as self-induction (368), which is an *E.M.F.* affecting the current on opening and closing the circuit. The ends of the coils are connected to the heavy brass blocks, C^1, C^2, C^3, on top of the box. When a brass plug, as P, is inserted between two blocks, as C^2 and C^3, the current practically all goes through the plug; but when the plug is withdrawn, as between C^1 and C^2, the current must go through the corresponding coil.

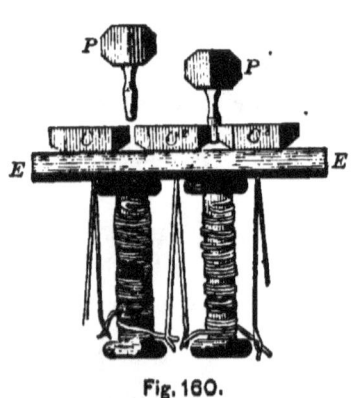

Fig. 160.

The resistance of these coils is often arranged in ohms of the following numbers:

1, .2, 2, 5, 10, 10, 20, 50, 100, 100, 200, 500, etc. In this way any number of ohms from one up to the full capacity of the resistance box can be thrown into circuit by withdrawing the proper plugs.

362. Divided Circuits.—When the wire leading from

any electric generator is divided into two branches, as at B (Fig. 161), the current also divides, part flowing by one path and a part by the other. The sum of these two currents is always equal to the current in the undivided

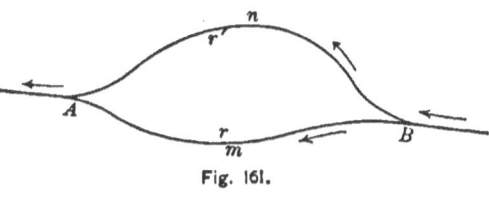

Fig. 161.

part of the circuit, since there is no accumulation of electricity at any point. Either of the branches between B and A is called a *Shunt* to the other, and the currents through them are inversely proportional to their resistances.

363. Resistance of a Divided Circuit.— Let the total resistance between the points A and B be represented by R, that of the branch BmA by r, and of BnA by r'. The conductivity of BA equals the sum of the conductivities of the two branches; and as conductivity is the reciprocal of resistance, then the conductivities of BA, BmA, and BnA are $\frac{1}{R}$, $\frac{1}{r}$, and $\frac{1}{r'}$ respectively, and $\frac{1}{R} = \frac{1}{r} + \frac{1}{r'}$. From this we easily derive $R = \frac{rr'}{r+r'}$. To illustrate, let a galvanometer whose resistance is 100 ohms have its binding-posts connected by a shunt of 50 ohms resistance; then the total resistance of this divided circuit is $\frac{100 \times 50}{100 + 50} = 33\frac{1}{3}$ ohms. The introduction of a shunt lessens the resistance between the points connected, as A and B.

364. Fall of Potential along a Conductor. — When a current flows through a conductor a difference of potential exists, in general, between different points on that conductor. Let A, B, C be three points on a conductor conveying a current, and let there be *no source of E.M.F. between these points*. Then if the current flows from A to B, the potential at A is higher than at B, and the potential at B is higher than at C. If the potential difference between A and B and that between B and C be measured, the ratio of the two will be the same as the ratio of the resistances between the same points. This is only another statement of Ohm's law. For since $C = \dfrac{E}{R}$, and the current is the same through the two adjacent sections of the conductor, the ratio of the effective E.M.F.'s to the resistances of the two sections is the same, since E.M.F.'s in this case mean potential differences. This important principle, of which great use is made in electrical measurements, may be expressed by saying that the *fall of potential along a conductor is proportional to the resistance passed over.*

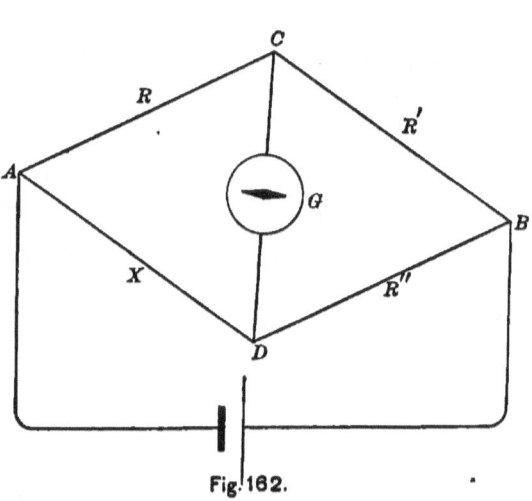

Fig. 162.

365. The Wheatstone Bridge consists of four resistances connected as shown in Fig. 162. The four conductors R, R', R'', X constitute the *Arms*, and the conductor CD the *Bridge*. On closing the circuit the current divides at B, the parts reuniting at A. The fall of potential along BCA is the same as that along BDA. Now if no current flows through the galvanometer, then the potentials of C and D must be the same. Under these conditions the fall of potential from B to C is the same as from B to D. We may get an expression for these two potential differences, and place them equal to each other. Let I' be the current through R'; it will also be the current through R, because none flows across through the galvanometer. Also let I'' be the current through the branch BDA. Then the difference of potential between B and C is the same as between B and D, and by Ohm's law (353)

$$R'I' = R''I''. \quad \ldots \ldots \quad (a)$$

In the same way, the equal potential difference between CA and DA gives

$$RI' = XI''. \quad \ldots \ldots \quad (b)$$

Dividing (b) by (a), we have

$$\frac{R}{R'} = \frac{X}{R''}.$$

When no current passes through the galvanometer, the four resistances in the arms of the bridge form a proportion. If three of them are known the proportion gives the formula, $X = \dfrac{RR''}{R'}$.

EXERCISES.

1. How shall several chromic acid cells be connected in order to heat as hot as possible a piece of platinum wire? Why?

2. If a circuit has a large resistance, would a long-coil or a short-coil galvanometer show the greater deflection? Why?

3. If an electro-magnet has a helix of a few turns of heavy copper wire, how would you join four chromic acid cells in order to make the magnet the strongest possible?

4. What facts indicate an intimate relation between the electric current and magnetism?

5. In the process of electrotyping, why must the article to be plated be connected to the negative electrode?

6. A single Daniell cell will not electrolyze acidulated water, but two joined in series will. Explain.

7. Roget suspended a spiral of fine elastic wire with its lower end just touching a dish of mercury. By connecting the poles of a battery to the upper end of the wire and to the mercury respectively the spiral vibrated up and down, producing bright sparks at the point of contact of the spiral with the mercury. Explain.

8. What will be the resistance of a mile of copper wire whose diameter is 0.032 inch?

9. How many feet of German-silver wire whose diameter is 0.01 inch will be needed for a resistance coil of 1000 ohms?

10. On measuring the resistance of a piece of copper wire 18.12 yds. long, it was found to be 3.02 ohms; what must be the length of another piece of the same diameter to have a resistance of 22.65 ohms?

11. If an incandescent lamp of 80 ohms resistance takes a current of 0.75 ampere, what $E.M.F.$ is required to light it?

12. A battery consists of 5 Daniell cells, each having an $E.M.F.$ of 1.1 volts and an internal resistance of 4 ohms; what current will they furnish when in parallel circuit through an external resistance of 7 ohms?

13. How many cells in series, each of 1.8 volts $E.M.F.$ and 1.1 ohms internal resistance, will be required to send a current of 0.5 ampere through an external resistance of 50 ohms?

14. With an external resistance of 9 ohms a certain battery gives a current of 0.43 ampere; when the external resistance is increased to 32 ohms the current falls to 0.2 ampere. What is the resistance of the battery?

15. How much silver will be deposited in one hour by a cell whose $E.M.F.$ is 1.8 volts, the total resistance of the circuit being 2.2 ohms?

16. A galvanometer whose resistance is 50 ohms is to be reduced to one of 5 ohms by a shunt. Compute the resistance of the shunt.

XV. CURRENT INDUCTION.

366. Currents induced by Magnets. — Exp. — Place a coil of insulated wire in circuit with a sensitive galvanometer (Fig. 163). Thrust suddenly into the helix the N-seeking pole of a strong bar

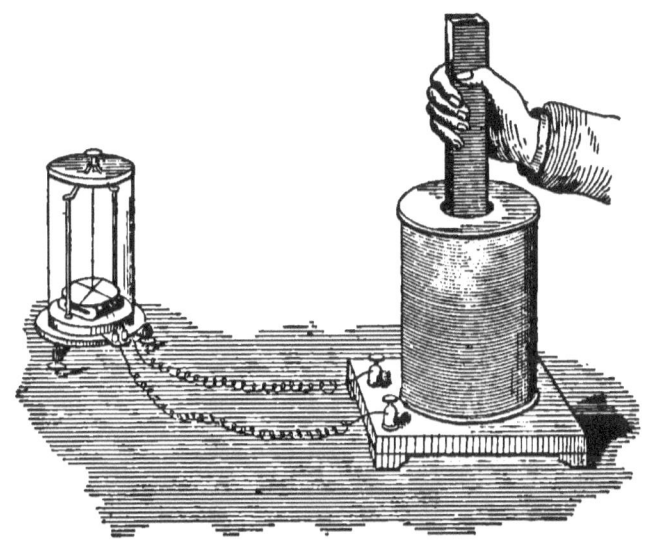

Fig. 163.

magnet. The needle will be deflected in a direction showing that a momentary current has been started in the coil opposite to that in which the hands of a watch move. On removing the magnet a current is produced in the opposite direction. If the S-seeking pole be used, results the reverse of these are obtained. If we substitute a helix of a smaller number of turns, or a magnet of less strength, the deflection is less marked, showing that a weaker current is set flowing, and hence a smaller $E.M.F.$ is set up. If the rate of approach or recession be less, the current produced is less.

The momentary currents set up in the helix are known

as *Induced Currents*. When the relative positions of a magnet and a closed conductor are so altered that a continuous variation is produced in the number of the lines of force threading through the conductor, induced currents are set up in the conductor in accordance with the following laws:

I. *An increase in the number of lines of force threading through a helix produces an indirect current, while a decrease in the number of lines produces a direct current.* By direct current is meant one flowing in the direction of the motion of watch-hands, and by indirect current, one flow-

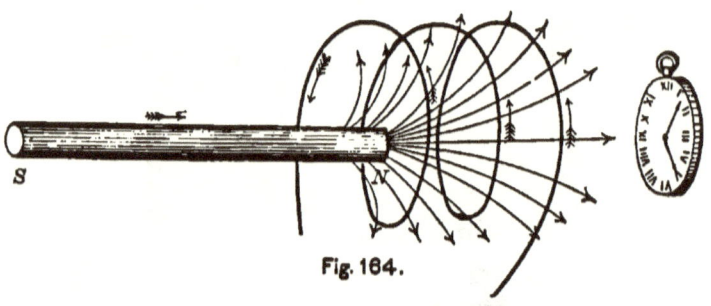

Fig. 164.

ing in the opposite direction. The observer must always be looking along the lines of force. Thus, if in Fig. 164 the magnet be thrust into the coil in the direction of the arrow, the current will flow counter clock-wise, as shown by the arrows on the coil. The motion of the magnet into the coil carries its lines of force with it, and thereby increases the number passing through the coil.

II. *The E.M.F. produced is equal to the rate of increase or decrease in the number of lines of force threading through the helix.*

Exp. — Wind a number of turns of fine insulated wire around the armature of a horseshoe magnet, leaving the ends of the iron free to come in contact with the poles of the permanent magnet. Connect the ends of the coil to a sensitive galvanometer, the armature being in contact with the magnetic poles, as shown in Fig. 165. Keeping the magnet fixed, suddenly pull off the armature. The galvanometer will show a momentary current. Suddenly bring the armature up to the poles of the magnet; another momentary current in the reverse direction will flow through the circuit.

With the armature in contact with the poles of the magnet, the number of lines of force passing through the coil is a maximum. When it is pulled away, the number of

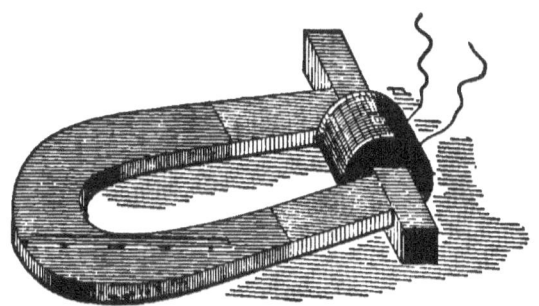

Fig. 165.

magnetic lines threading through the coil rapidly diminishes. This illustrates Faraday's method of obtaining electric currents by the agency of magnetism.

367. Currents induced by Other Currents. — **Exp.** — Insert within a helix connected to a sensitive galvanometer a second helix in circuit with a battery (Fig. 166). The effects are exactly like those obtained by the use of the magnet, even to the direction of the galvanometer deflection, when the polarity of the helix (344) corresponds with that of the magnet. Furthermore, the effects are the same if the helix is inserted before closing the circuit.

The experiment shows that *the approach of a current to*

a closed conductor, or its recession from it, has the same effect in inducing a current as a magnet. This was to be expected, since we have seen that a helix carrying a cur-

Fig. 166.

rent has properties identical with a magnet. The coil connected with the battery is called the *Primary Coil*, and the other the *Secondary Coil*.

368. The Extra Current or Self-Induction. — Exp. — With a battery, a long helix of wire with an iron core, and a suitable galvanometer, set up a circuit as shown in Fig. 167. On closing the circuit, the needle is deflected by a part of the current passing through the coil of the galvanometer. Now bring the needle back to zero, and place some obstacle in its way so that it

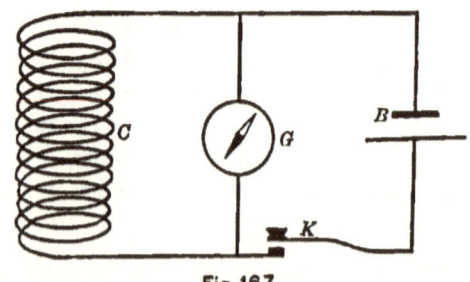

Fig. 167.

cannot move in that direction, but is free to swing in the opposite. Break the circuit at K; the needle will be deflected, showing a momentary current through the galvanometer in an opposite direction to the battery current.

This current is due to an $E.M.F.$ of self-induction generated by the inductive action of the several turns of the coil on one another at the moment of opening the circuit. It flows through C in the same direction as the battery current, but completes the circuit through the galvanometer instead of the battery, and hence reverses its deflection. It is often called the *Extra Current, but should always be recognized as a current of self-induction.* A study of Fig. 168 may make its origin clearer. Let A and B represent adjacent convolutions of the helix. On closing the circuit some of the lines of force which belong to one coil thread through the other, thereby increasing the number. Hence, each will induce an opposing $E.M.F.$ in the other (366). On breaking the circuit, the reverse will be the case. The $E.M.F.$ generated is proportional to the current strength and to the number of convolutions of the helix; in other words, it is proportional to the number of lines running through the coil, and is therefore increased by adding an iron core (342).

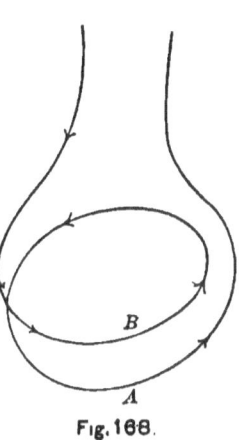

Fig. 168.

369. **The Induction Coil** is an apparatus for producing currents of very high $E.M.F.$ It consists of a helix of coarse insulated wire surrounding an iron core, and this again surrounded by a second helix of a large number of

turns of fine wire (Fig. 169). The inner or primary coil is connected to a battery through a current interrupter; also through a device called a *Commutator* for changing the direction of the current. At the "make" and "break" of the circuit currents are induced in the secondary coil in accordance with the laws of current induction (366).

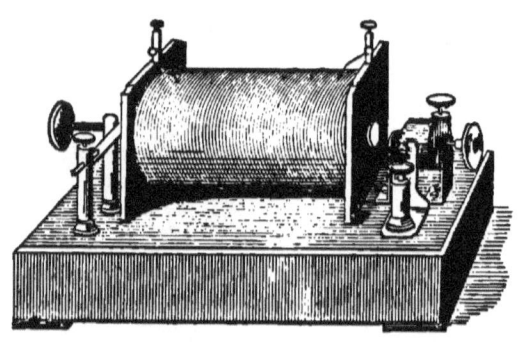
Fig. 169.

In coils designed to give sparks between the poles of the secondary coil a *Condenser* is added. This is placed in the supporting base of the coil, and consists of two sets of interlaid layers of tin-foil, separated by paper saturated with paraffin. These two sets are connected to two points of the primary circuit on opposite sides of the current interrupter.

370. Action of Coil.—Let Fig. 170 represent an induction coil, showing the arrangement of the various parts. The current entering at m ascends the spring O, crosses to the spring H, which carries the soft iron I', then along q to the primary coil P, around which it circulates, making the right-hand end the N-seeking pole of the helix; and finally back through l to the battery. Its iron core I becomes a magnet and attracts to itself I', thereby breaking the circuit by separating H from the horizontal screw. At the instant the current begins to circulate in P a current is induced in the secondary coil flowing counter clock-wise, that is, from S' to S through the wire (366). When the current is broken in the primary a momentary current from S to S' is induced. Hence, momentary alternating currents are induced in the secondary coil by the intermittent current in the primary. On account of the great number of coils of wire composing the secondary

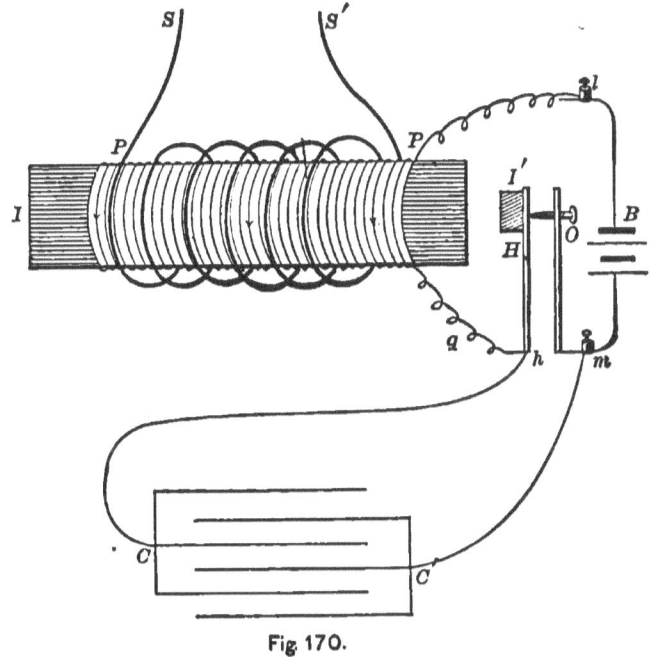

Fig. 170.

coil a very high $E.M.F.$ is set up in it. This is increased by the action of the core in increasing the number of lines of force which pass through the entire helix.

The self-induction of the primary has a very important bearing on the action of the coil. At the instant the circuit is made, the counter $E.M.F.$ opposes the battery current, and thereby prolongs the time of reaching its greatest strength. Consequently the $E.M.F.$ of the secondary coil will be diminished by self-induction in the primary. The $E.M.F.$ of self-induction at the "break" of the primary is direct, and this added to the $E.M.F.$ of the battery produces a spark at the break points H, O. For an instant the primary current may actually be increased at the break of the circuit.

371. The Condenser. — The addition of a condenser increases the $E.M.F.$ of the secondary coil in two ways: 1. It gives such an increase of capacity to the primary coil that at the moment of breaking the circuit the potential difference between the points H, O does not rise high enough to break across the air-gap. The interruption of the primary is therefore more abrupt, and the $E.M.F.$ of the secondary is increased. 2. After the break, the surface C' of the condenser being of higher potential than C, and connected with it through the battery, the condenser discharges back through the battery and the primary coil, thereby sending a reverse current, which materially aids in demagnetizing the core. Rapid demagnetization of the core means great rate of change in the lines of force threading through the secondary, and hence high $E.M.F.$ On breaking the circuit, the direct induced $E.M.F.$ in the secondary is therefore much higher than the inverse induced $E.M.F.$ on closing the circuit.

372. Experiments with the Induction Coil:

1. *Physiological Effects.* — Hold in the hands the electrodes of a very small induction coil, such as are used by physicians. On closing the battery circuit a peculiar muscular contraction is produced.

The "shock" from large coils is very dangerous, owing to the high $E.M.F.$ generated. The physiological action of such shocks is not fully understood. It appears that the danger decreases with the increase in rapidity of the impulses or alternations. Recent experiments with induction coils, worked by alternating currents of very high frequency, have demonstrated that the discharge of the secondary may be taken through the body without injury.

2. *Mechanical Effects.* — Hold a piece of cardboard between the electrodes of a coil throwing a spark an inch long. The card will be perforated, leaving a burr on each side. Thin plates of any non-conductor can be perforated in like manner.

3. *Heating Effects.* — Make a torpedo (306) and place it in circuit with an induction coil. On closing the battery circuit the powder will be exploded.

4. *Chemical Effects.* — Place on a plate of glass a strip of white blotting-paper moistened with a solution of potassium iodide and starch paste. Attach one of the electrodes of a small induction coil to the margin of the paper. Handle the other electrode with an insulator and trace characters with it on the paper when the coil is in action. The discharge decomposes the chemical salt, as shown by the blue mark. This blue mark is due to the reaction between the iodine and the starch.

5. *Luminous Effects.* — Exhaust the air from the *Aurora Tube* (Fig. 171), and connect its extremities to the poles of the induction coil and observe its appearance in a dark room. If the tube is well exhausted and a little vapor of alcohol or naphtha is afterwards introduced into the tube the light will be beautifully stratified.

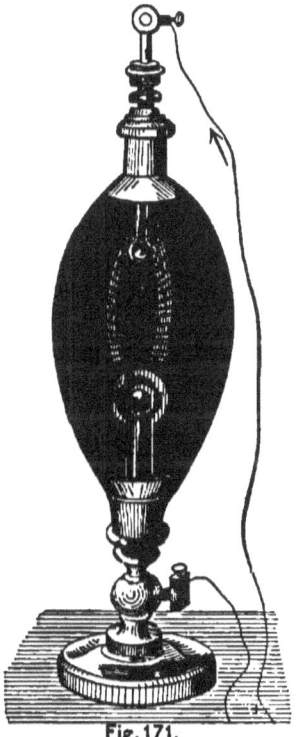

Fig. 171.

Coat a vase or glass goblet with tin-foil, a little over half-way up inside, and place it on the table of the air-pump, beneath a bell-jar provided with a brass sliding-rod passing air-tight through the cap at the top (Fig. 172). Surround the part of the rod inside the jar with a glass tube and push the tube down till it touches the tin-foil. Connect the rod and the air-pump table to the electrodes of the induction coil. On exhausting the air a beautiful play of light will fill the bell-jar. If the vase is made of uranium glass, or if it stands on a block of that material, the effect is still more beautiful. This experiment, known as *Gassiot's Cascade*, should be conducted in a darkened room.

Fig. 172.

Geissler Tubes (Fig. 173) are glass tubes of various patterns containing highly rarefied vapor or gas. Platinum wires are fused into the glass for the admission of the current. Exceedingly beautiful effects are obtained by placing such a tube in circuit with the induction coil.

6. *The intermittent character of the discharge* is demonstrated by the multiple images of any moving object in a dark room near a luminous Geissler tube; for instance, the hand moved to and fro over the tube gives the impression of several hands. The phenomenon is due to the intermittent character of the light, and the fact that images formed on the retina of the eye are retained for a finite length of time.

Fig. 173.

XVI. THE DYNAMO.

373. **The Dynamo** is a machine for converting the energy of mechanical motion into the energy of an electric current. The so-called generation of electricity consists always in the generation or production of an *E.M.F.* or of electric pressure. The quantity of electricity at our command is as definite and invariable as the quantity of energy. No battery, dynamo, or other device creates

electricity. It creates electromotive force, by means of which electricity may be accumulated on an insulated conductor and in a condenser; or by means of which electricity may be made to flow through conducting circuits.

In the conducting circuit, external to the region where the electric pressure is applied, the electricity flows from a higher electric level or potential to a lower, as water flows from a higher elevation to a lower. Within that part of the circuit where the *E.M.F.* originates the electricity is forced from a lower electric level to a higher, as

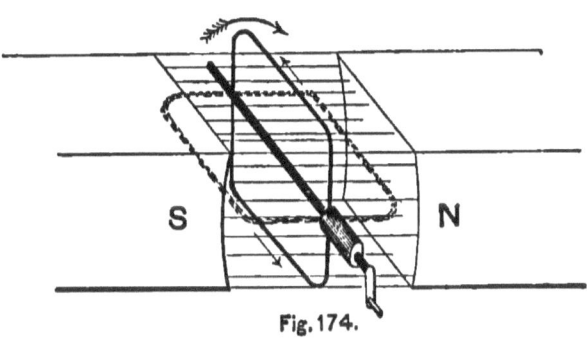

Fig. 174.

water is pumped from a lower to a higher level. In a voltaic cell this latter region appears to be at the crossing of the current from the zinc to the liquid; in the dynamo machine it is in that part of the machine called the armature, which usually revolves between the poles of a powerful electro-magnet.

The dynamo is based on the principles of current induction. It contains a system of closed conductors, revolving in a magnetic field in such a way as to vary continuously the number of lines of force threading through them.

374. The Ideal Simple Dynamo. — Suppose a single

loop of wire to revolve between the poles of a magnet NS (Fig. 174) in the direction of the arrow and round a horizontal line as an axis. The lines of force run across from N to S, as indicated by the light lines. The loop of wire in the position shown encloses the largest possible number of lines of magnetic force. When it has revolved through 90°, or a quarter of a turn, the lines of force will be parallel to its plane and none will thread through it. During this quarter turn, the number of lines has been decreasing and consequently a direct current has been produced, looking from N toward S, as shown by the arrows. During the next quarter turn the lines will increase again, but will run

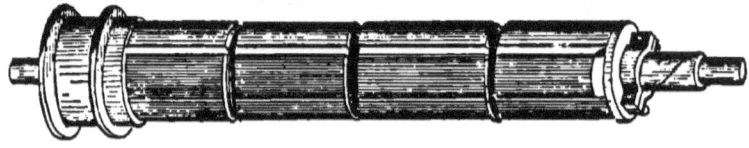

Fig. 175.

through from the opposite side of the loop. Hence, the current will continue to flow in the same direction round the loop. During the next half revolution the current round the loop will flow in the opposite direction. The current through the loop reverses twice, therefore, in each revolution.

375. The Shuttle Armature.— If the number of turns of wire composing the revolving coil be increased from one to many, the $E.M.F.$ generated by the revolution of the coil will be increased in the same ratio. If, further, the "magnetic field" between the poles N and S contain iron, the number of lines of force will be augmented (257). These results were secured first by Siemens in the *Shuttle Armature* (Fig. 175), of which Fig. 176 is a cross-section. Two broad grooves are ploughed in an iron cylinder on opposite sides, and in these grooves are wound the spirals or loops of wire. This form of armature is not suitable for large currents, since currents are produced in the iron core as well as in the insulated wire; and these absorb energy and heat the iron. Moreover, the current falls to zero every half revolution, and this is very undesirable for operations requiring steady currents. This armature, however, is extensively used in the "magneto bells" used for "calling up" on telephone circuits.

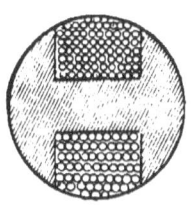

Fig. 176.

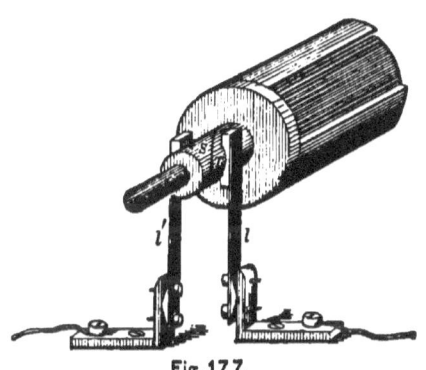

Fig. 177.

376. The Commutator. — In order to have the successive oppositely-directed currents flow in the same direction through the external circuit, a special device must be employed called a *Commutator*. The shuttle armature employs a two-part commutator, shown in Figs. 177 and 178. It consists of a split tube with the two halves insulated from each other, and from the shaft on which they are mounted. The two ends of the coil on the armature are connected with the two halves of the tube. Two brushes bear on the commutator, and they are so placed that they exchange connections with the commutator segments at the same time that the current

Fig. 178.

reverses through the coil. In this way one of the brushes is kept always positive and the other negative, and the current flows from the positive brush through the external circuit back to the negative brush, and thence through the armature to the starting point.

377. The Field-Magnet. — In dynamo machines the magnetic field is produced by a large electro-magnet excited by the current flowing from the armature, which is led either wholly or in part round the field-magnet cores. When the entire current is carried round the coils of the field-magnet the dynamo is said to be a *Series Dynamo*. Such a one is shown diagrammatically in Fig. 179. When the field-magnet is excited by coils of many turns of fine wire connected as a shunt to the external circuit, the dynamo is said to be *Shunt Wound*. The connections of the field and external circuit are shown in Fig. 180. A combination of these two methods of exciting the field-magnet is called *Compound Winding*. The residual magnetism remaining in the cores is sufficient to start the machine. The currents thus produced increase the magnetism and so increase the $E.M.F.$

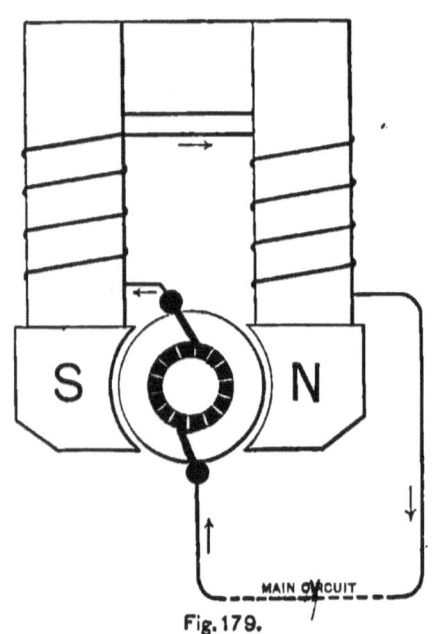
Fig. 179.

378. The Drum Armature. — The modern *Drum*

Armature (Fig. 181) is a direct outgrowth of the shuttle armature. Instead of a single coil wound in a groove, it has a good many coils or sections wound on the outside of a cylindrical iron core, made of thin disks of soft iron insulated from one another. A single section is wound on, the armature is then turned through a small angle, and another section is wound, and so on till the entire circumference has been gone over. The sections are then all joined together in series, and the junctions of the sections are connected to bars, insulated from one another, on the commutator. It will readily be seen that when the brushes bear on opposite bars of the commutator, the current has two paths through the armature, and divides equally between them. The binding wires round the outside hold the coils in place. With this winding and with numerous sections, sometimes as high as ninety-six or even more, the current is almost perfectly continuous.

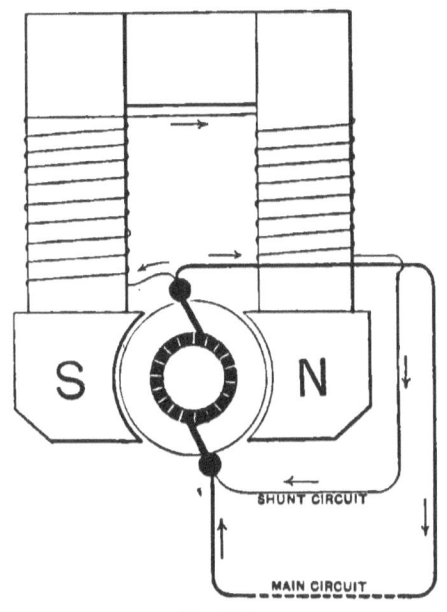

Fig. 180.

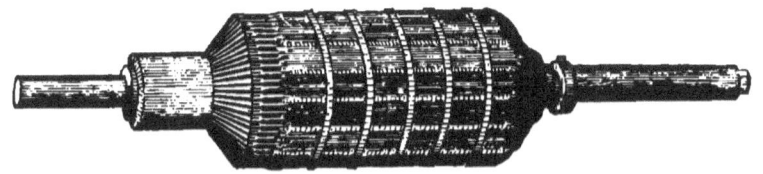

Fig. 181.

XVII. THE ELECTRIC LIGHT.

379. Electric Lights are of two classes, *Arc* and *Incandescent*. The former is produced by a current of from six to ten amperes passing across from one carbon rod to another through the highly heated vapor as a conductor. The carbon rods are placed in electrical connection with the two poles of a dynamo, and the current is established by

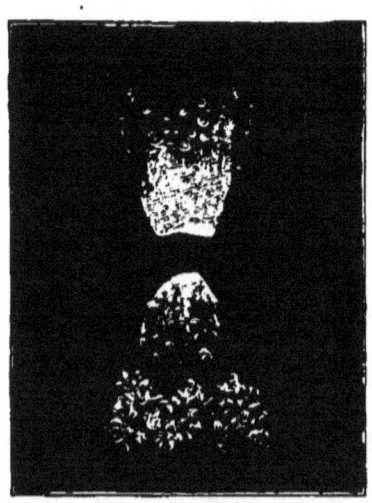

Fig. 182.

bringing the carbon points together. Upon separating them they are heated to an exceedingly high temperature and the current continues to pass across through the heated vapor. The ends of the carbon rods are disintegrated, a depression or crater forming in the positive (Fig. 182) and a cone on the negative carbon. Most of the light of the electric arc comes from the bottom of the crater, which is the hottest part of the carbon. Sir Humphry Davy, who first produced the electric arc light, obtained an arc between two charcoal points four inches in length. For this purpose he used a great battery of 2000 cells. The pores of the charcoal sticks had been filled with mercury. Despretz, with 600 Bunsen cells, obtained an arc 7.8 inches long. The arc light may be produced in a vacuum. The intense heat is not due, therefore, to combustion, but is due to the conversion of the energy of the current into heat by means of the back $E.M.F.$ of the arc. The electric arc light may be produced even in water, but its brilliancy is much reduced and the water is rapidly decomposed.

380. **The Adjustment of the Carbons.**— The distance between the ends of the carbon rods for a ten-ampere current, and a potential difference between them of 50 volts, is about one-eighth of an inch. As the carbons wear away they must be made to approach each other automatically. For the purpose of keeping the carbons at the required distance the constant-current arc lamp is provided with two electro-magnets, one of coarse wire to carry the entire current through the lamp, and the other of fine wire and of about 100 ohms resistance, placed as a shunt around the arc. A small part of the current then passes by way of this shunt or by-path, the strength of the shunt current depending upon the difference of potential between the two carbons. As the carbons burn away and the arc lengthens this potential difference increases and more current goes through the shunt. The apparatus is so made that the shunt magnet operates to make the carbons approach each other when the current through it reaches a predetermined value, which is adjustable by means of a spring. The shunt magnet accomplishes this result either by removing a *detent* so as to allow a train of wheels to drop the upper carbon, or it releases a *clutch*, which normally

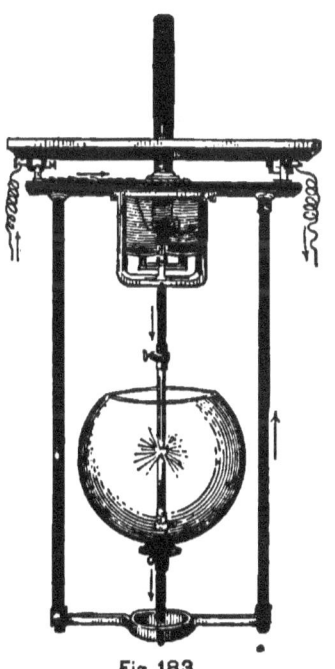

Fig. 183.

grasps the carbon holder, and which then allows it to slide through. The shortening of the arc reduces the potential difference between the carbons, and the shunt magnet immediately ceases to operate. This process is repeated as soon as the carbon rods wear away again to the feeding point. Fig. 183 shows one form of the Brush arc lamp.

Arc lamps to the number of 125 are sometimes placed in series, requiring an *E.M.F.* of at least 6500 volts to operate them.

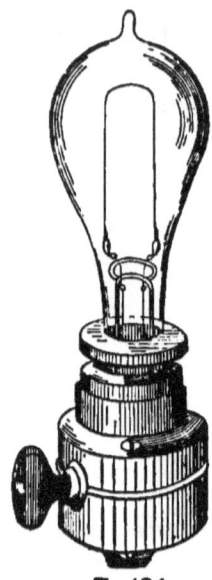

Fig. 184.

381. Incandescent Lamps. — The smaller subdivision of the electric light is made by means of carbon filaments enclosed in exhausted glass bulbs (Fig. 184). These small lamps are usually placed in parallel across from one main conductor to the other (Fig. 185). The carbon filaments have a resistance of from 25 to 200 ohms when hot, according to the voltage used and the candle power of the

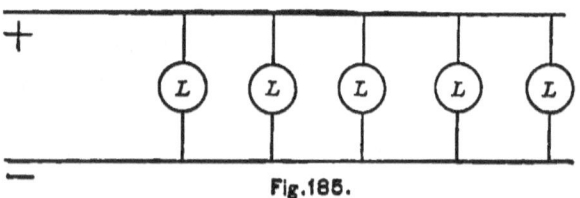

Fig. 185.

lamps. On account of this resistance the filament becomes heated to high incandescence. No combustion takes place, but the carbon gradually deteriorates and is thrown off so as to blacken the bulbs. The lamp then gives less light, and if burned long enough, the carbon filament will break and interrupt the current.

XVIII. ELECTRIC MOTOR AND TELEGRAPH.

382. An Electric Motor does mechanical work at the expense of electric energy. It is similar to a dynamo. When an electric current is sent through the armature, the reaction between the magnetic field and the currents flowing across it (317) causes the armature to turn. The power applied in watts is the product of the current in amperes and the potential difference between its terminals in volts.

383. The Telegraph is a device for transmitting signals through the agency of an electric current. Its essential parts are the *Line*, the *Transmitter or Key*, the *Receiver or Sounder*, and the *Battery*.

384. The Line is a heavy wire, iron or copper, insulated from the earth, except at the ends, and serving to connect the signalling apparatus. The ends of this conductor are connected to large metallic plates buried in the earth. By this means the earth is made a part of the circuit of the electric generator employed.

385. The Transmitter or Key (Fig. 186) is merely a current interrupter, and usually consists of a brass lever *A*, turning about an axle at *B*. It is connected to the line by the screws *C* and *D*. On pressing the lever down, a platinum point

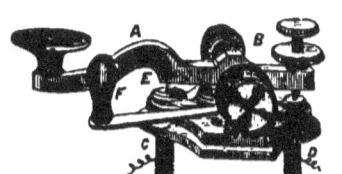

Fig. 186.

projecting under the lever is brought in contact with another platinum point E, thereby closing the circuit. When not in use, the circuit is left closed, the switch F being used for that purpose.

386. The Receiver or Sounder (Fig. 187) consists of

Fig. 187.

an electro-magnet A with a pivoted armature B. When the circuit is closed through the terminals D and E, the armature is attracted to the magnet, producing a clicking sound. On breaking the circuit, a spring C pulls the armature away from the electro-magnet.

387. The Relay. — When the resistance of the line is large, the current is not likely to be strong enough to operate the sounder with sufficient energy to render the signals distinctly audible. To remedy this, an electro-magnet, called a *Relay* (Fig. 188), whose helix A is composed of many turns of fine wire, is placed in the circuit by

Fig. 188.

means of its terminals C and D. As its armature H moves to and fro between the points at K, it opens and closes a second and shorter circuit through E and F, in which the sounder is placed. Thus the weak current, through the agency of the relay, brings into action a strong current which does the necessary work.

388. The Battery consists of a large number of cells, usually of the gravity type, connected in series. (Why?) It is generally divided into two sections, one placed at each terminal station, these sections being also connected in series through the line.

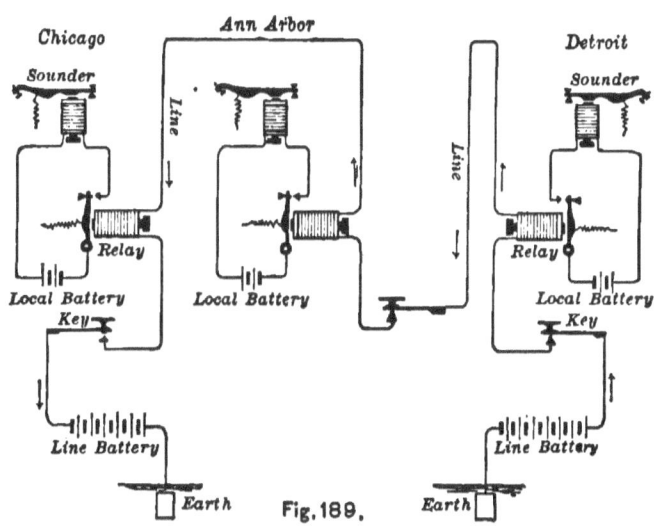

Fig. 189.

389. The Signals are a series of armature clicks separated by intervals of silence of greater or less duration, a short interval between the clicks being known as a "dot," and a long one as a "dash." By a combination of "dots" and "dashes" letters are represented and words are spelled out.

390. The Telegraph System described in the preceding articles is known as Morse's, from its inventor.

Fig. 189 illustrates diagrammatically the instruments necessary for three stations, together with the mode of connection.

391. **The Electric Bell** (Fig. 190) consists of an electro-magnet E, one end of whose helix is connected with the binding-post P, and the other end is connected, through the spring R and the screw N, to the post W. These posts are connected by wire to the poles of a battery. In this circuit is a *Push Button B*, or some other form of circuit-closer.

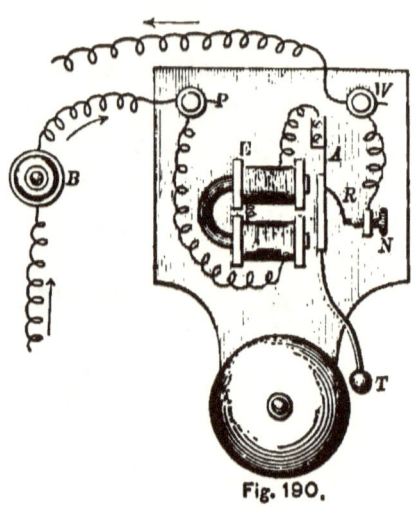

Fig. 190.

A cross-section of the push button is shown in Fig. 191. On closing the circuit, the electro-magnet draws the pivoted armature to it, thereby causing the ball T to strike the gong. The contact between the spring R and the screw N is now broken, the armature is released, the spring A carrying it back till R and N are in contact, when the circuit is again closed, and the action is repeated.

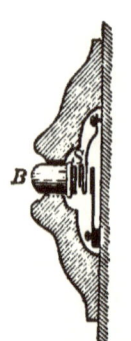

Fig. 191.

XIX. THE TELEPHONE AND MICROPHONE.

392. **The Bell Telephone** (Fig. 192) consists of a steel magnet C, one end of which is inserted in a helix of fine copper wire b, whose ends are connected to the

binding-posts t and t. At right angles to the magnet, and not quite touching the pole within the helix, is an elastic diaphragm or disk a of soft sheet-iron, kept in place by the conical mouth-piece d. If the instrument is placed in an electric circuit when the current is unsteady, or alternating in direction, the magnetic field due to the helix, when combined with that due to the magnet, alters intermittently the number of lines of force which branch out from the pole, thereby varying the attraction of the magnet for the disk. The result is that the disk vibrates in exact keeping with the changes in the current.

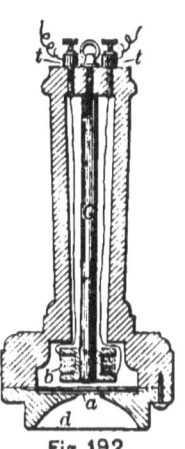

Fig. 192.

393. **The Microphone** is a device for rendering an electric current unsteady by means of a variable resistance in the circuit. One of its simplest forms is shown in Fig. 193. It consists of a rod of gas-carbon A, whose tapering ends rest loosely in conical depressions made in bars C, C, of the same material attached to a sounding-board B. These bars are placed in circuit with a battery and a Bell telephone. While the current is passing, the least motion of the sounding-board, caused either by sound-waves or by any other means, moves the loose carbon pencil and varies the pressure between its ends and the supporting bars. Since a slight increase of pressure between two conductors when resting loosely one on the

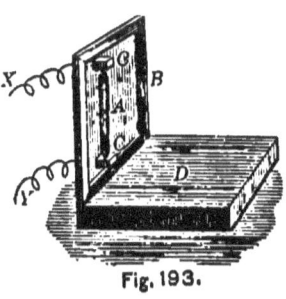

Fig. 193.

258 ELEMENTS OF PHYSICS.

other lessens the resistance of the contact and conversely, then the vibrations of the sounding-board cause variations in the pressure at the points of contact of the carbons and consequently corresponding fluctuations in the current and vibrations of the telephone disk.

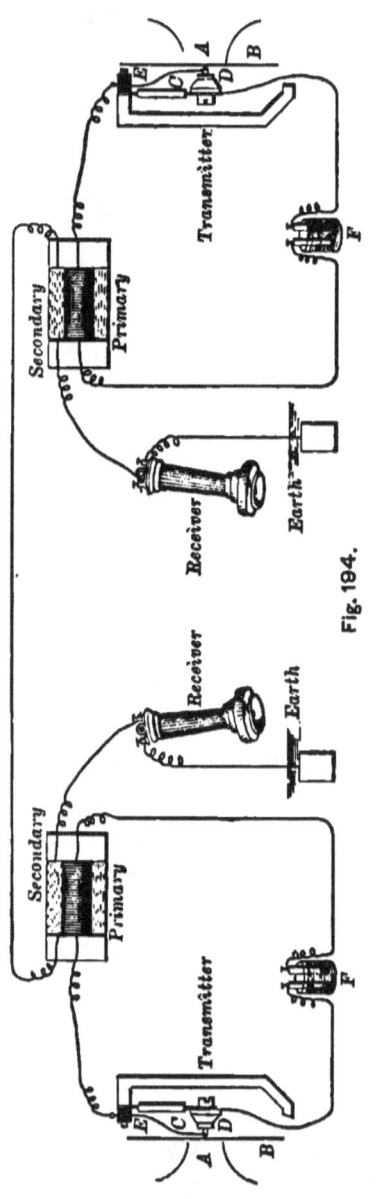

Fig. 194.

394. The Blake Transmitter is a form of microphone used in connection with the Bell telephone in the telephone service of our cities and towns. An idea of its construction and mode of action can be got from an examination of Fig. 194, which illustrates diagrammatically two places connected by telephone. A conical mouthpiece A has back of it a diaphragm of elastic sheet-iron secured by its edges to the supporting frame. Back of this disk are two springs E and C, insulated from each other at the top. The spring C has a carbon button D attached to its lower end, against which rests the light hammer-shaped end of the other spring. These springs are connected respectively to the poles of a battery, through the primary coil of a small induction coil. One of the poles of the secondary coil is connected to the earth; the other pole is connected to the line wire leading to the other station. In circuit with this secondary coil is a Bell telephone.

Sound-waves striking against the diaphragm A cause a varying pressure between the free ends of the springs E and C, which act as a microphone. The battery current is consequently unsteady, and capable of inducing currents in the secondary coil (367). These induced currents, although of higher $E.M.F.$, will yet have all the peculiarities of the inducing current. The telephone in this secondary circuit will be affected by these induced currents, and will accordingly send off sound-waves similar to those which disturbed the disk of the transmitter.

If a telephone is used as a transmitter as well as a receiver, its action is as follows: Sound-waves striking the diaphragm set it in vibration, and cause alternating currents to flow through its helix and the external circuit (343, 366). These currents operate the receiving telephones. No battery is used in this arrangement.

CHAPTER VI.

SOUND.

I. WAVE-MOTION.

395. Vibration. — **Exp.** — Suspend a ball by a long thread and set it swinging like a pendulum. Notice that the ball returns periodically to the starting point. Now set the ball moving in a circle, the string describing a conical surface. The ball again returns periodically to the starting point.

A Vibrating Body is one whose parts are moving in such a manner that each returns periodically to its initial position. *A Complete Vibration*, or simply *a vibration*, is the motion comprised between two successive passages of the object in the same direction through any position.

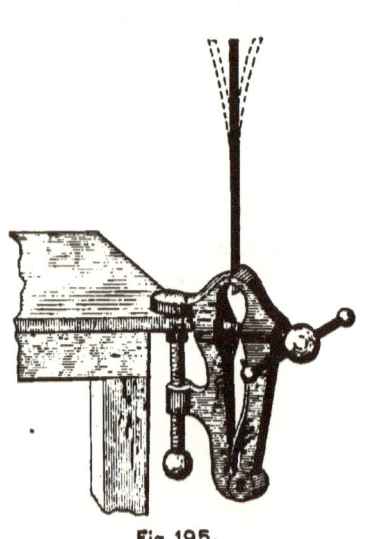

Fig. 195.

396. Vibrations classified. — **Exp.** — Clamp one end of a strip of lath in a vise (Fig. 195). Draw the free end aside and then release it. It flies back, passing through its original position to a point nearly as far on the opposite side. Each point of the lath executes a vibratory movement in a path at right angles to its length.

Vibrations of this character are called *Transverse Vibrations*.

SOUND. 261

Exp. — Fasten one end of a long spiral spring[1] to a hook in the wall and hold the other end in the hand. Insert a finger-nail between two turns of the wire, a short distance from the hand, and pull them asunder. On releasing the coils, they are set swinging to and fro in line with the length of the spiral. Fig. 196 shows the appearance of a portion of the spiral after releasing the coils.

Fig. 196.

Vibrations of this character are called *Longitudinal Vibrations*.

Exp. — Rigidly fasten a heavy ball to a fine wire about one metre long and support it like a pendulum. Turn the ball part way around on its axis. On releasing it, it returns periodically to its position of rest, as the wire alternately twists and untwists.

Vibrations of this character are called *Torsional Vibrations*.

397. The Conical Pendulum. — **Exp.** — Suspend a ball by a long thread. Set it swinging in a circle and count how many vibrations it makes in some period of time, as 30 sec. Place the eye on a level with the ball and it will appear to move to and fro in a straight line, increasing its speed from either end to the middle and decreasing its speed as it recedes from the middle. Now set it swinging in a small arc like a common pendulum, and count how many complete vibrations it makes in the previously selected period of time. The two pendulums will be found to have the same period. If the eye were placed below the common pendulum, the ball would appear to move to and fro in a straight line, changing its speed much like the conical one.

398. Locating the Vibrating Body. — Let the circle

---

[1] Such a spring may be made by winding iron or brass wire, No. 18, on a long rod.

whose centre is D (Fig. 197) represent the circular path of the conical pendulum bob, and AG its apparent linear path. Let b, c, d, etc., mark the positions of the ball at equal intervals of time. Then B, C, D, etc., will mark the corresponding apparent places on the line AG.

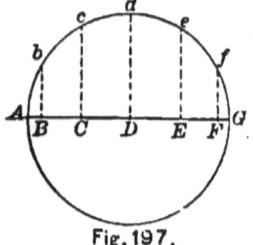

Fig. 197.

399. Simple Harmonic Motion. — A body moving to and fro along a straight line in the manner that the pendulum bob appears to do is said to have *Simple Harmonic Motion*. The distance DA or DG is the *Amplitude of Vibration*. The *Period of Vibration* is the time-interval between two successive passages of the body through any point in the same direction.

400. Waves. — Exp. — Lay straight on the floor a soft cotton rope, about 5 feet long. Holding the other end of the rope in the hand, set up vibrations in the rope by a quick movement of the hand up and down. Each point of the rope will be seen to vibrate transversely with simple harmonic motion, and the disturbance started by the hand will move along the rope.

The curved forms which are seen traversing the rope are called *Waves;* these may, in general, be defined as the configuration of a body caused by its parts vibrating, and passing successively through corresponding positions.

401. Graphic Representation of Waves. — Let a number of particles A, B, C, etc., at equal distances apart, move to and fro in parallel straight lines, with simple harmonic motion, each particle beginning its journey at $\frac{1}{16}$th of the period behind the one to the right. Let the problem be to locate these moving points at some moment of time and determine the wave-form.

Draw the circle O (Fig. 198) with a radius equal to the amplitude of vibration. The paths AA', BB', CC', etc., of the particles will each be the apparent path of a particle moving uniformly in the circumference of this circle. The particle at 1 will appear at A; the one at 2, being $\frac{1}{16}$th of a period in advance, will appear at b; the one at 3 will appear at c; and so on for d, e, f, etc. If we pass a smooth curved line through these points, we have a picture of a wave-motion like that seen in the rope. An

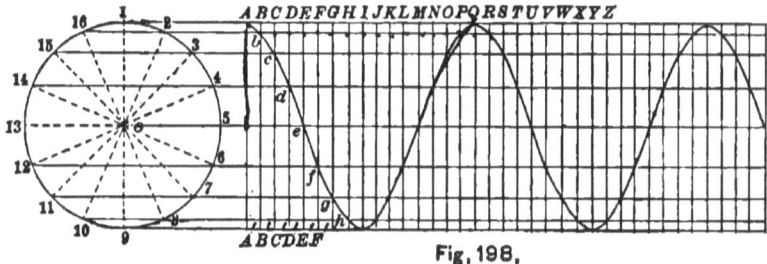

Fig. 198.

inspection of the figure shows that this wave-form is the result of compounding, at right angles, a simple harmonic motion with a uniform linear motion.

The following experiment will assist in making this subject clear:

Clamp a strip of lath by one end in a vise. To the other end attach a lead pencil, the apparatus to be arranged so that the point of the pencil just touches a sheet of cardboard resting on the table. Set the bar in vibration and slowly draw the cardboard under the pencil in a line parallel to the bar. The pencil will trace a wave-like line on the paper. A camel's hair brush and ink may be used instead of the pencil.

402. Wave-Length. — *The Length* of a wave is the

distance from any particle in the wave to another particle similarly placed; as from A to Q, D to T, F to V, etc. An inspection of Fig. 198 shows that while the wave-form travels from A to Q, or Q to A, particle A makes a complete vibration. Hence, *the wave-length is the distance traversed by the wave during one vibration-period.*

403. Waves on a Medium. — **Exp.** — Drop a pebble into a vessel of water. Circular waves will be seen moving outward from the disturbed point to the sides of the vessel. A small cork floating on the surface merely rises and falls with the movement of the water and is not carried along with the wave.

The waves on the surface of a liquid are due to its par-

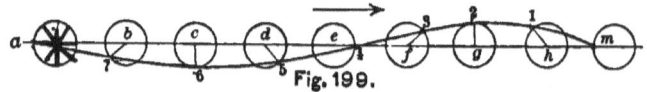

Fig. 199.

ticles moving in closed curves, circular when the amplitude is small. That such a motion will produce a wave-form is shown by Fig. 199. Each circle is divided into eight equal parts; a particle is supposed to move in the circumference of each circle at the same rate, but in any two consecutive circles to be at points separated by $\frac{1}{8}$th of a period. Then when a has completed one revolution, b will be $\frac{1}{8}$th behind, c will be $\frac{2}{8}$ths behind, etc. A smooth curve traced through the points thus located represents the form of the surface of the water.

Waves of this kind on liquids and in ropes are known as *Waves of Troughs* and *Crests*.

404. Waves through a Medium. — **Exp.** — Place on the table a tin tube about 3 metres long and 10 cm. in diameter, one end

SOUND. 265

of which tapers to a diameter of about 2.5 cm. (Fig. 200). Tie over the large end a paper membrane, and in front of the small end place the flame of a lighted candle. Tap the membrane with the finger; the candle is visibly agitated and perhaps extinguished.

On tapping the membrane, it is made to move forward suddenly. In so doing it compresses the air imme-

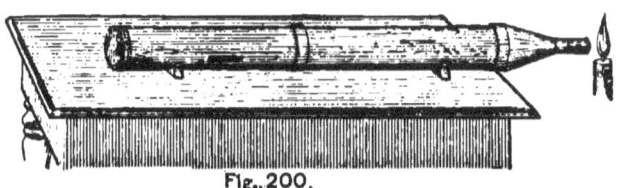
Fig. 200.

diately in front of it. These air particles are forced apart by their elasticity and compress the air further along the tube. The continued repetition of this process carries the disturbance through the tube to the flame. Each air particle vibrates to and fro in a short path in line with the direction of the propagation of the disturbance. (What kind of a vibration is this?) The arrangement of the air particles at the time of propagation may be represented by the lines of Fig. 201.

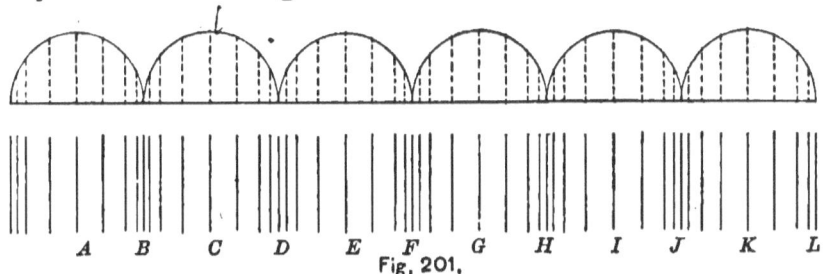
Fig. 201.

A study of the figure shows that the particles are separated by more than the average distance at the points A, C, E, etc., and by less than the average distance at B, D, F, etc., these letters marking the places of greatest rarefaction and condensation respectively. Particles at A and C are said to be in the *same phase*, because they are in the same stage of vibration. The distance AC is a *wave-length* and comprises a rarefaction and a condensation. Such waves are usually spoken of as *Waves of Condensation and Rarefaction*.

405. Velocity of Waves. — **Exp.** — Stretch side by side two similar rubber tubes, about 3 m. long, giving them equal tensions. Strike the tubes with a ruler, a short distance from the end, causing a depression in each. These depressions will move along each tube to the other end, where they will be reflected as elevations, returning together to the end first disturbed. Now increase the tension of one of the tubes. Again start at the same time a depression in each tube. It will travel much faster along the tube of greater tension. Stretch a third tube filled with sand beside the other two. A wave along it will travel more slowly than along the others.

Hence, it appears that the velocity of the propagation of a transverse wave-motion along a stretched body depends on its tension and on its mass per unit of length (451). For longitudinal vibrations the velocity depends on the coefficient of elasticity and the density, or $v = \sqrt{\dfrac{e}{d}}$.

EXERCISES.

1. Account for the waves which are often seen moving across fields of grain.
2. On what does the length of a wave depend?
3. How could you prove that there is no motion of translation in water waves?

II. SOURCES OF SOUND.

406. Sound, as distinguished from hearing, consists of those vibratory motions of matter which are capable of affecting the organs of hearing.

SOUND.

407. Its Origin. — **Exp.** — Suspend a small ball so as to touch the edge of a large bell-jar. Strike the edge of the jar with a cork mallet. The ball will be repeatedly thrown away from the jar so long as the sound is heard. What must be the condition of the jar?

Exp. — Stretch a steel wire between two screw-eyes in a board. Draw a violin-bow across the wire, and then touch it with a light ball suspended by a thread. So long as the wire emits sound, the ball will continue to be thrown away from the wire. Inference?

Exp. — Tap a tuning-fork (Fig. 202) against a block of soft wood, and while sounding, touch one prong to the surface of water. In what condition is the fork shown to be?

Exp. — Insert a short wooden or tin whistle in one end of a glass tube about 40 cm. long and 2 cm. in diameter (Fig. 203). Close the other end with a cork, having distributed evenly within the tube a little fine dry cork-dust, obtained by filing a cork. Hold the tube in a horizontal position, and blow the whistle. Notice the agitation of the cork-dust, and that it is deposited in little piles transverse to the axis of the tube. What must be the condition of the air in the tube while the whistle

Fig. 202.

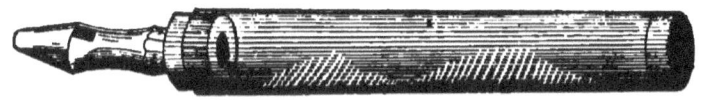

Fig 203.

is sounding? Does the effect on the cork-dust suggest a vibratory motion of the air particles, or one of translation?

These experiments clearly teach that sound-emitting bodies are vibrating bodies, and that where there is no sound there is either no motion, or the motion is too feeble to affect the organs of hearing. Bodies emitting sounds are called *Sonorous Bodies*.

III. TRANSMISSION OF SOUND.

408. The Question Stated. — If sound is caused by a vibrating body, the question arises, How is the energy of this vibratory movement transmitted to the ear? Must there be a material medium connecting the sounding body with the ear, and must this medium have any special properties?

409. Air a Medium. — **Exp.** — Place an alarm-clock on a thick bed of cotton under a bell-jar on the air-pump. Set the alarm to ring in a few minutes. Exhaust as completely as possible.

If the vacuum is good, the sound of the alarm may be only faintly audible, or it may not even be heard at all. The experiment shows that air transmits sound, and that sound does not traverse a vacuum.

410. Any Gas a Medium. — **Exp.** — Fit to a large, heavy, wide-mouthed bottle a rubber stopper with two holes in it. Pass a glass rod through one of the holes, and attach to it as large a bell as will enter the mouth of the bottle. Through the other opening insert a glass tube, and connect it by a rubber tube to the air-pump. Exhaust the air as completely as possible from the bottle. Close the rubber tube with a pinch-cock, disconnect the tube from the pump and connect it to a supply of hydrogen. After the gas is introduced, the sound of the bell can be distinctly heard. Other gases can be tested in a similar manner.

Hence, it appears that any gas can serve as a medium for sound.

411. Liquids as Media. — **Exp.** — Fill a tumbler nearly full of water and place it on an empty chalk-box. Stick the stem of a tuning-fork into a small disk of wood. Set the fork in vibration and hold it with the wooden disk resting on the water in the tumbler.

SOUND. 269

The sound of the fork will be heard as if coming from the box, having been transmitted to it by the water. If any other liquid be used in the tumbler, the result will be much the same.

We conclude from this that liquids are good conductors of sound.

412. Solids as Media. – **Exp.** – Place a vibrating tuning-fork against one end of a long wooden rod. Hold the other end firmly against the door of the room. The sound of the fork will seem to come from the door, the vibrations having been transmitted to it by the rod. If a thick layer of cotton-wool be placed between the door and the rod the sound is scarcely audible.

Wood is highly elastic, but the cotton-wool is not; hence it appears that solids transmit sound provided they are elastic.

413. The Acoustic Telephone is a familiar instance of a solid transmitting sounds. It consists of a string or wire, of any desired length, attached to the thin elastic bottoms of two small boxes. If the string be properly stretched, any sound produced in the one box is easily heard by applying the ear to the other.

Solids are the best conductors of sound. The report of a cannon has been heard more than 250 miles by applying the ear to the solid earth. The great eruption of Cotopaxi in 1744 was heard distinctly over 500 miles, although several gigantic mountains and numerous deep valleys intervened.

414. Sound-Waves. —When a body, as a tuning-fork, is set in vibration, disturbances, known as *Sound-Waves*, are produced in the air around it. These waves consist of a series of condensations and rarefactions, succeeding each other at regular intervals, and forming concentric spherical shells of air of different densities. Each air particle swings to and fro in a very short path along the radius of the sphere, that is, the vibrations are longitudinal.

IV. REFLECTION AND REFRACTION OF SOUND.

415. Reflection of Waves. — Exp. — Connect the bottoms of two empty chalk-boxes by the spiral spring of Art. 396, gently stretching the spiral. Start a pulse through it by pressing together a few of its turns at one end and suddenly releasing them. The wave moves rapidly along the spiral till it reaches the other end; then it returns, as if reflected, to the starting point, and continues travelling to and fro till the energy is dissipated.

The experiment demonstrates that a wave-motion can be reflected from a surface.

416. Reflected Sound-Waves. — Exp. — Place a large concave reflector, such as is used behind wall lamps, where the light of the sun can strike it squarely and ascertain where the light reflected from its surface is brought most nearly to a point. That point will be the focus (500) of the reflector. Now suspend a loud-ticking watch a few inches in front of this focus, as at W (Fig. 204), and observe that by standing at a certain distance, as E, in front of the reflector you can hear the watch distinctly; but if the reflector be removed, the ticking is nearly, if not quite, inaudible.

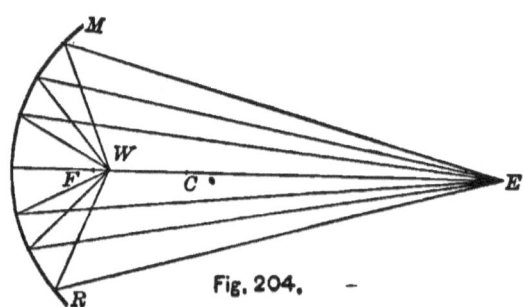

Fig. 204.

The sound-waves striking against the curved surface are reflected to a point in a manner quite similar to light-waves (501).

417. **An Echo** is the repetition of a sound due to its reflection from some distant surface, as that of a building, cliff, cloud, body of water, etc. The time-interval between producing a sound and hearing its echo is the time that sound takes to travel from the source to the reflecting body and thence to the ear. The sensation of sound occupies about $\frac{1}{10}$th of a second, and during that time the sound-wave travels about 110 feet (429). If then the reflecting surface be distant 55 feet, a sound having a duration of $\frac{1}{10}$th of a second will be followed immediately by an echo, since the first sound-wave will have just time to travel to the reflecting surface and back before the cessation of the original sound. Multiple or repeated echoes are due either to independent reflections from bodies at different distances, or to successive reflections, as in the case of parallel walls at a suitable distance apart. The roll of thunder is partly due to multiple reflection.

If the distance of the reflecting body be too small to produce a distinct echo, then the reflected sound will tend to render the original sound indistinct, as is the case in large empty buildings; or it will tend to strengthen the original sound, as illustrated by the greater ease with which a speaker can be heard in a room of medium size.

418. **A Whispering Gallery** is a room where a whisper or other faint sound produced at some particular point is heard distinctly at some distant part of the building. The phenomenon is due to one of two causes: First, the walls of the room may be either elliptical or circular in form, and hence act like concave reflectors in converging all the sound-waves to one point; or, secondly, the sound may be reflected repeatedly from point to point, and thus be made to travel round the walls. This is the case

with long, circular passage-ways, which are somewhat like speaking-tubes.

419. Refraction. — **Exp.** — Fill a large rubber toy balloon with carbon dioxide. Suspend a loud-ticking watch from a support, and place yourself at a point where you can hear it fairly. Now hold the inflated balloon between the watch and your ear. By moving it slowly back and forth you will find a position where the ticking is much louder.

This is due to the fact that the globe of dense gas acts on sound-waves like a convex lens on light-waves (525), converging them to a focus. If the balloon be filled with hydrogen gas, an opposite effect will be produced, the sound being diffused instead of being brought to a focus, similar to the action of a concave lens on light (526).

The refraction of sound is caused by the change in velocity suffered by that part of the sound-wave which enters the globe of gas, a subject which will be more fully considered under refraction of light (516).

EXERCISES.

1. Why is it that any sound produced in a forest is followed by several similar sounds coming from as many different directions?
2. The report of a gun returns to the gunner in 2¾ seconds; how far away is the reflecting surface?
3. Why are there echoes in a very large empty room? Why do they disappear when filled with people?
4. Why does a partially deaf person place his open hand back of his ear when listening to a speaker?
5. Why is it that the wind blowing against a series of sound-waves deflects them upward, so that the sound of a distant bell is often audible at the top of a tower, when it is inaudible at the base.

V. INTENSITY AND LOUDNESS.

420. The Physical Intensity of a sound depends on the energy of the vibrating particles of the transmitting medium; whereas the *Loudness* of a sound depends not only on the energy of the vibratory movement, but also on the condition of the ear in which the sensation is produced. The rate of vibration also plays an important part in estimating loudness, as the ear is not equally sensitive to all sounds.

421. Effect of Amplitude. — Exp. — Set a tuning-fork in vibration by striking one of its prongs a feeble blow with a block of wood. The sound emitted is faint, and the prongs when touched to water do not throw it with much energy. If you strike it a sharp blow, the sound is much louder, and the prongs throw water on touching it with marked vehemence.

To understand the above, it should be remembered that sonorous bodies, like pendulums, make their vibrations in equal intervals of time. In the second case, the amplitude of vibration is greater than in the first; consequently, the velocity of the prongs is greater, and also that of the disturbed air particles. Since the energy of these particles varies as the square of their velocities, *the intensity of the sound must vary as the square of the amplitude.*

422. Effect of Density of Medium. — Exp. — Place an alarm-clock under a bell-jar on the air-pump table. As the air gradually exhausted from the bell-jar, the sound of the clock grows fainter.

Exp. — Fill a large bell-jar with hydrogen. As with one hand you hold it up,[1] introduce in it with the other a small bell. The sound is

---

[1] When the gas is hydrogen, the jar must be held mouth downward; but when filled with carbon dioxide, the reverse is the case.

much feebler than when the jar is filled with air. On repeating the experiment with the jar full of carbon dioxide, the sound is much louder than in air.

These differences of loudness are due to differences in the energy of the wave-motions produced in the gases. Since energy depends on mass, the energy of the wave-motion set up in a rarefied or light gas is less than that in a dense one; and since the energy of the vibratory body is absorbed less rapidly by the rare medium, the body continues in vibration for a longer time. On high mountains, where the air is quite rare, conversation is carried on with difficulty, and the report of a gun is no louder than the noise due to the breaking of a small twig under ordinary circumstances.

423. Effect of Distance. — As the sound-waves move outward from the vibrating body, each spherical layer of air imparts its energy to the enveloping one. Since these layers are surfaces of spheres, the number of particles composing them increases as the squares of their radii. Hence, the energy of the individual particles must decrease in like ratio, that is, *the intensity of sound varies inversely as the square of the distance from the sonorous body.*

During the vibratory movement of the air, heat is developed through friction among the air particles, and some of the mechanical energy is transformed into heat. The decrease in the intensity of sound is therefore considerably greater than that given by the theoretical law of inverse squares, on account of the energy dissipated as heat.

424. The Speaking-Tube. — The weakening of sound from the enlargement of the sound-waves as they recede from the sonorous body would evidently not take place if the sonorous body were within a tube. Under such a condition the sound-waves would not be propagated as concentric spheres, but the successive layers of air affected would be of equal mass, and the sound would be conveyed with but little loss of intensity. Tubes used in this way are called *Speaking-Tubes*.

425. Effect of Area of Vibrating Body. — Exp. — Compare the sound of a small tuning-fork with that of a large one of the same pitch (441). The large one produces the louder sound.

In order that a vibrating body may be a sonorous one, the condensations and rarefactions produced in the surrounding medium must be well marked. When the object is small, the particles of the medium can readily flow around it and thus prevent the formation of clearly defined waves. On the other hand, when it is large, this is not possible, and consequently the condensations and rarefactions are more pronounced, thereby making the sound louder. Hence, *the intensity of sound depends on the area of the sonorous body.*

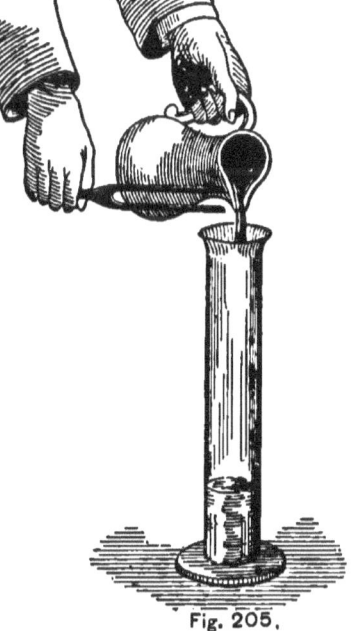

Fig. 205.

Illustrations of this fact are found in many stringed musical instruments, where the sound is intensified by placing two or more strings side by side when they are fine; and, secondly, by placing a sounding-board beneath them to be set in motion by the string. The loud sound produced by many wind instruments is due to the fact that the air within the broad aperture opposite the mouth-piece is a vibrating body of large area.

426. The Resonator. — **Exp.** — Hold a vibrating tuning-fork over the mouth of a cylindrical jar, of about 2 inches diameter and 12 inches deep (Fig. 205). Pour in water slowly, and notice that as the air-column grows shorter the sound grows louder till a certain length is obtained, after which it grows weaker. If forks of different pitch are tried, each is found to have its own length of air-column which reënforces its sound.

This strengthening of the sound is called *Resonance*, and any vessel enclosing a body of air producing this effect is called a *Resonator*.

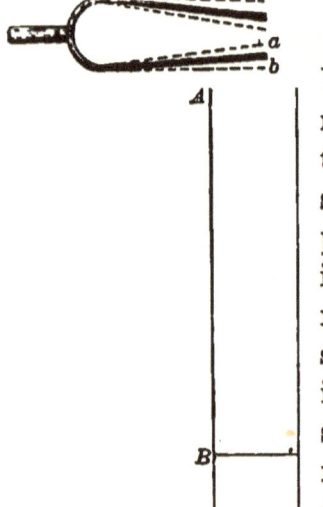

Fig. 206.

427. Its Action Explained. — When the prong a (Fig. 206) moves to b it makes half a vibration, and hence generates half a sound-wave. The condensation it produces passes down the tube AB, is reflected from the bottom, and returns to unite with other waves sent out by the prong. Now, if AB is of the proper length, this condensation can move down the tube and return to combine with the condensation produced by the prong moving from b to a, thus making the condensation more marked, and thereby strengthening the sound. The effect of the fork on the column of air is to set it in vibration, and the layer of air at its mouth has the sound-producing properties of a sonorous body of large area.

If the tube is lengthened or shortened a trifle, the reflected condensations and rarefactions do not issue at the mouth at the proper time to combine with those of the same phase produced by the fork in the air outside the tube, and consequently they act against each other, causing an enfeeblement of the sound (421).

428. Length of Air-Column. — The fork in swinging from a to b makes one-half of a vibration. During that time, if the condensation travels from A to B and back again, then AB must be one-fourth of a wave-length. If AB be three-fourths of a wave-length, then the first condensation which passes down the tube will return in time to unite with the one produced by the prong on its second swing from b to a. Any odd multiple of the one-fourth of a wave-length will reënforce the sound. (Why?)

It must be borne in mind that both theory and experiment show that the length of the air-column which will reënforce a sound is affected by the diameter of the column. The column shortens on increasing the width, the true length being given in the case of cylindrical columns by increasing the length by about two-fifths of the diameter.

VI. VELOCITY OF SOUND.

429. History of the Problem. — It has long been known that sound is not propagated instantaneously. The first reliable experiments made to determine the velocity in air were those of a commission of the French Academy appointed in 1738. In 1822 another company of physicists undertook a series of observations to solve this problem. The method of procedure was to divide into two parties, each provided with a good chronometer and a small cannon. Each party recorded the interval between the observed flash of the cannon and the report. An average of a large number of observations gave as a result 1118.152 feet per second at a temperature of 16° C. The defect in the method is that the perceptions of sound and of light are not equally quick, and vary with different persons. Stone determined the velocity of sound in 1871 by stationing two observers three miles apart to give signals by electricity on hearing the report of a cannon. This method employs the sense of hearing only. The result was 332.4 metres, or about 1090 feet at a temperature of 0° C.

430. Velocity in Gases. — Since the velocity of propagation of a wave through any medium varies directly as the square root of the coefficient of elasticity and inversely as the square root of the density (405), it follows that sound travels four times as fast through hydrogen as through oxygen under the same pressure, the density of oxygen being 16. If a gas be subjected to pressure, its elasticity and density will change at the same rate, and consequently the velocity of sound in the gas will not be affected. Since heating a gas increases its coefficient of elasticity, it will also increase the velocity of sound. Experiment and theory show that the correction is 0.6 metre, or nearly two feet, for each degree centigrade.

431. Velocity in Liquids. — Colladon and Sturm in 1827 conducted a series of experiments in the water of Lake Geneva, and found that sound travels in water at the rate of 1435 metres per second at 8° C. Subsequent

SOUND. 279

experiments made at higher temperatures show that the velocity is considerably affected by changes of temperature. Such great velocity is due to the high elasticity of water as compared with its density.

432. **Velocity in Solids.** — Solids, like liquids, have a high coefficient of elasticity, and accordingly transmit sound with great rapidity. In iron, sound has a velocity of 16,820 feet per second; in glass, 14,850 feet per second; in lead, a metal of low elasticity, only 4,030 feet per second.

EXERCISES.

1. The flash of a cannon was seen, and ten seconds later the report was heard; how far off was the cannon, the temperature being 70° F.?
2. A stone dropped into the mouth of a mine was heard to strike the bottom in two seconds; how deep was the mine?
3. Why is the echo weaker than the original sound?
4. Does sound travel faster at the foot of a mountain than at the top?
5. What is the velocity of sound in coal gas at 0° C., the specific gravity being 0.5 that of air?
6. What must be the temperature of air in order that sound may travel in it with the same velocity as in hydrogen at 0° C.?

VII. INTERFERENCE.

433. **Combining Waves.** — Exp. — Fasten one end of a soft cotton rope, about 20 feet long, to a suitable support. Holding the other end in the hand, start a wave, or pulse, in it by a quick movement up and down. The wave will travel along the rope as an elevation or crest, and on reaching the fixed end will be reflected, returning to the hand as a depression, or trough. At the moment the wave is reflected at the far end, start a second wave. These will meet near the middle of the rope. The returning wave will tend to move this point downward and the

advancing wave to carry it upward. If these forces are equal, the point will remain at rest; but if not, its motion will be proportional to the difference.

The experiment illustrates this principle: *If two waves pass simultaneously through the same medium, the*

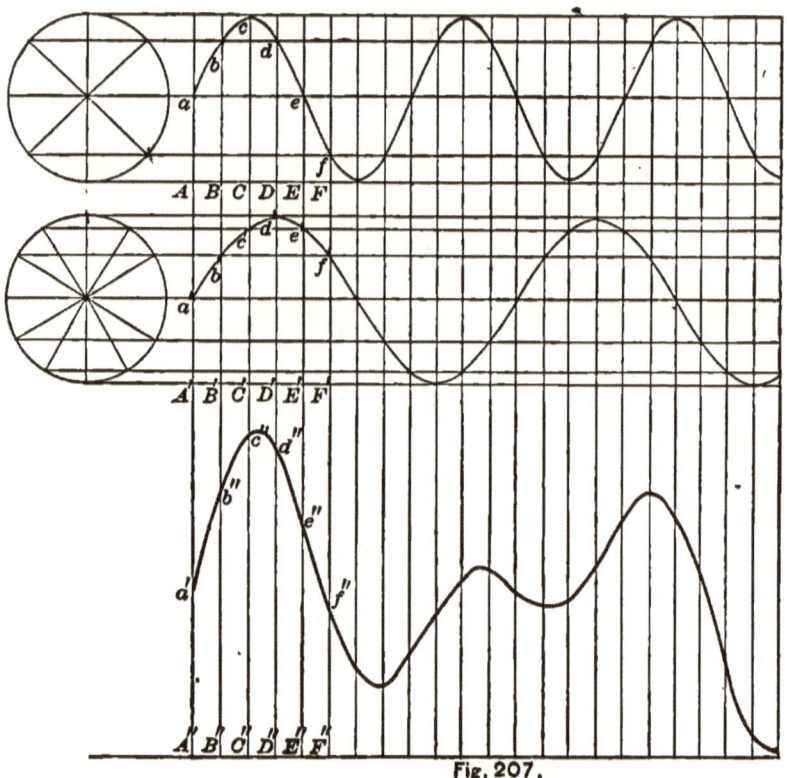

Fig. 207.

actual motion of each particle is the resultant of the motions due to each system separately; it equals their sum when the directions of motion are the same, and their difference when they are opposite.

Let the first two curved lines (Fig. 207) represent two wave-motions in the same medium, having the same amplitude, but differing in wave-length. The heavy line will represent the resultant wave, the points being determined in the following manner: $A''a'' = A'a' + Aa$, $B''b'' = B'b' + Bb$, $C''c'' = C'c' + Cc$, etc. A study of the figure shows that sometimes the two waves act together, resulting in an increased amplitude, while at other times the motions impressed on the particles are opposite in direction, thereby reducing the amplitude.

434. Interference is the result of two wave-motions traversing a medium at the same time. If two sound-waves of equal amplitude meet in opposite phases, the result will be an extinction of sound. This is illustrated by the following experiment:

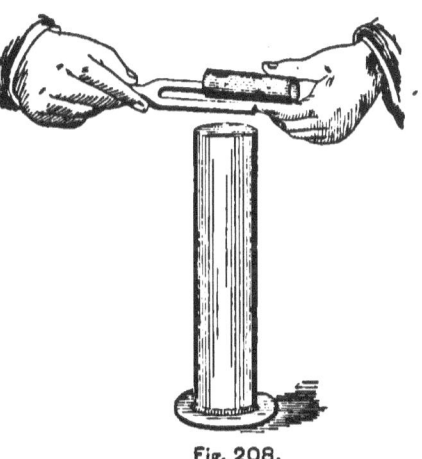

Fig. 208.

Exp. — Hold a vibrating tuning-fork over a cylindrical jar, serving as a resonator. Turn it slowly round its axis, and notice that in certain positions it is nearly inaudible. When in one of these positions slide a small paper cylinder over one prong (Fig. 208), being careful not to touch it. The sound will be restored.

Each prong sends a set of waves down the jar. When the fork is held with one corner of a prong over the mouth of the jar, the two sets of waves will not reach the bottom in the same phase. Hence, the two wave-motions act against each other, producing either partial or total extinction of sound. The correctness of this explanation is shown by the restoration of the sound when the waves from one prong are cut off by the paper cylinder.

435. Beats. — **Exp.** — Select two large tuning-forks giving the same tone. When the forks are set in vibration, the sound is smooth as if only one fork were sounding. Fasten a piece of wax to a prong of one of the forks; the sound will be marked by a peculiar palpitation or unsteadiness.

These outbursts of sound, followed by moments of comparative silence, are called *Beats*.

436. Number of Beats. — Let two sounds be produced by 100 and 120 vibrations per second, respectively. Then the first takes $\frac{1}{100}$th of a second for one vibration, and the other, $\frac{1}{120}$th of a second. When the first has made one vibration, the second has made as many as $\frac{1}{120}$ is contained in $\frac{1}{100}$, or $\frac{120}{100} = 1\frac{1}{5}$. Hence, it appears that the second gains $\frac{1}{5}$th of a period during each vibration of the first, and at the end of $2\frac{1}{2}$ vibrations on the part of the first, the second will be $\frac{1}{2}$ of a vibration in advance, that is, will be in exactly the opposite phase, producing rarefactions, while the first is producing condensations. This will evidently destroy the sound, if the amplitudes are equal, or weaken it, if they are unequal. During the next $2\frac{1}{2}$ vibrations, the second will gain another half vibration, and will be in the same phase as the first. For every 5 vibrations made by the first, there is then *one* beat, and $\frac{100}{5} = 20$ per second. But $120 - 100 = 20$. Therefore, the number of beats per second due to two sounds is equal to the difference of their vibration-rates.

VIII. SYMPATHETIC VIBRATIONS.

437. Forced and Sympathetic Vibrations. — **Exp.** — Construct two pendulums to vibrate in the same time, using a heavy ball and a light one for bobs. Suspend them from the same support and set the heavy one swinging; the other one will soon begin to vibrate. If you lengthen or shorten the light pendulum it will not so readily acquire the motion.

Exp. — Place on the table, a few feet apart, two tuning-forks of the same pitch. If you keep one of them in violent vibration for a short time you will discover, on stopping it, that the second fork is vibrating. If the experiment be repeated, with the pitch of one of the forks slightly changed by cementing a small weight to one prong, no such effect will be obtained.

These experiments prove that an elastic body can be thrown into vibration by another one whose vibrations have considerable energy, strongly when the normal periods of vibration of the two are the same, and feebly when they are not. The resulting vibrations are called, respectively, *Sympathetic* and *Forced Vibrations*.

438. **Explanation.** — When two bodies of the same vibration-rate are placed near each other and one of them is set vibrating, the motions of the first are communicated to the second by the material medium connecting them. The pulses in the medium reach the second body at intervals corresponding to its vibration-period, and their effect is cumulative, each impulse arriving just in time to add to the movement. If the two bodies do not vibrate in equal times, the impulses from the first will not agree in phase with the motions started in the second body and the disturbance can not accumulate.

439. **Illustrations.** — The sounding-board of a piano and the membrane of a banjo are *forced* into vibration by the strings stretched over them. The two prongs of a tuning-fork naturally vibrate at slightly different rates on account of unavoidable differences of size ; but being connected at the stem, the faster one tends to accelerate the slower one, and the slower one to retard the faster.

the result being that they agree in rate. Objects in a room often respond to certain notes of the piano, the effect being generally to cause a tingling sound as if there was a loose pin on the sounding-board.

IX. PITCH.

440. Musical Sounds are those which are pleasant to the ear, and are caused by regular periodic vibrations. A *Noise* is a disagreeable sound, either because of its short duration, as the report of a gun, or because the vibrations producing it are not periodic, or because it is a mixture of several discordant sounds (464), as the clapping of the hands.

Exp. — Attach a rose gas-burner to a metal pipe about 15 in. long, and connect it to the gas service with a rubber tube. Light the gas and notice the rustling sound attending its burning. Now hold a large tin tube, several feet long, over the burner. At a certain position of the flame within the tin tube, a sound like that of an organ-pipe will be obtained. With tubes of different lengths, the sounds will be of different pitch.

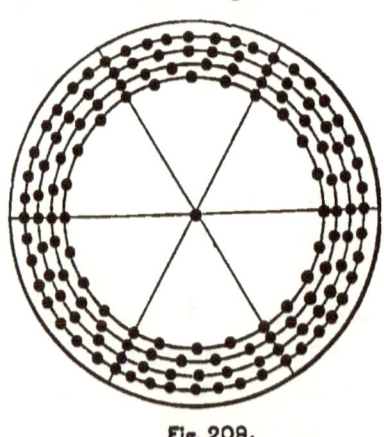

Fig. 209.

The experiment shows that the rustling of the flame is due to the mixing of many different sounds, the existence of which is proved by their reënforcement by the different air-columns.

441. Pitch. — **Exp.** — Cut a cardboard disk 15 cm. in diameter. Draw on it four concentric circles, having the diameters 13, 11.5, 10, and 8.5 cm. respectively. Divide these circles into 48, 36, 30, and 24 equal parts respectively (Fig. 209). At the points of division cut round holes .5 cm. in diameter. Mount the disk on the spindle of a whirling machine, and rotate it rapidly. With a glass tube direct a steady stream of air against one of the rows of holes. A different musical sound is produced for each circle.

Such an apparatus is called a *Siren*. The sounds are due to vibrations set up in the air by the regular succession of air-puffs passing through the holes as they cross the stream of air from the tube. The number of puffs per second will differ for the different circles of holes and also with the speed of the disk, thereby causing a difference in the number of sound-waves produced per second, or a difference of *Pitch*. The pitch of a musical sound is determined by the number of vibrations made by the body in a second. When the number of vibrations per second is very great, the pitch is *acute;* when small, *grave*.

442. Limits of Audibility. — The limits of perceptible pitch vary with the individual. According to Helmholtz, there must be at least 30 vibrations per second to produce a continuous sound, and when the number exceeds 38,000, it becomes inaudible. Other experimenters have placed the limits at 16 and 41,000. Most musical sounds are comprised between 27 and 4,000 vibrations per second.

443. Vibration-rate Measured. — The exact measurement of the vibration-rate of a sonorous body is a difficult matter. One of the simplest methods is by the use of the siren. The siren disk is given such a speed that the sound obtained is judged to be of the same pitch as the sonorous body in question. The number of revolutions made by the siren per second, multiplied by the number of holes in the circle used, will be the number of vibrations per second produced by the body.

444. Relation of Pitch to Wave-length.—If a tuning-fork makes 256 vibrations per second, and in that time a sound-wave travels 1,100 feet, then the first wave formed will be 1,100 feet from the fork, on the completion of the 256th vibration. Hence, in 1,100 feet there are 256 waves, and the length of each is $\frac{1100}{256}$ feet. Therefore, wave-length equals velocity of sound divided by vibration-rate, that is, $l = \frac{v}{n}$, in which v is velocity, l wave-length, and n vibration-rate.

445. Intervals.—When two notes differ in pitch they are said to be separated by an *Interval*, the value of which is obtained by finding the ratio of their vibration-rates. The most important interval is the *Octave*, the numerical value of which is 2; the next is the *Fifth*, the value of which is $\frac{3}{2}$; then follow the *Fourth*, the *Major Third*, and the *Minor Third*, equal to $\frac{4}{3}$, $\frac{5}{4}$, and $\frac{6}{5}$ respectively.

446. The Gamut or Diatonic Scale is a series of eight notes which succeed each other with gradually increasing pitch, the two extremes being an octave apart. The first, or lowest note, is called the *Key-Note*, and the last is regarded as the key-note of another set of eight notes. In this way the series is repeated till the limit of pitch is reached or a sufficiently extended scale is obtained. The tones comprised in each octave are named C, D, E, F, G, A, B, C'. The initial note of such a series may be given any pitch at pleasure. Physicists have agreed to assign to C (the "middle C" of a piano) 256 vibrations per second. In music there is no legal standard of pitch; that in general use in this country assigns to C 261 vibrations per second.

447. The Diatonic Scale accounted for. — It is found that when three notes whose vibration-numbers are as $4:5:6$ are sounded together, a pleasing effect is produced on the ear. Such a combination of notes is called a *Major Triad;* and together with the octave of the lowest of the three they constitute a *Major Chord.* The major diatonic scale is built on three sets of such chords. Representing by C the first of such a series, and assigning to it 256 vibrations, then the second will have $\frac{5}{4} \times 256 = 320$ vibrations, the third one $\frac{6}{4} \times 256 = 384$, and the fourth $2 \times 256 = 512$. These are called C, E, G, and C'. Now if we make G the first of another series of four, the second will have $\frac{5}{4} \times 384 = 480$ vibrations, the third $\frac{6}{4} \times 384 = 576$, or 288 when reduced by an octave, the fourth $2 \times 384 = 768$. The series now stands as follows:

256	288	320	384	480	512
C	D	E	G	B	C'

It is evident that the intervals from E to G and from G to B are much larger than that between any other two consecutive notes. If we make C' the third note of a major chord, we have for the first note $\frac{4}{6} \times 512 = 341\frac{1}{3}$, and for the second $\frac{5}{6} \times 512 = 426\frac{2}{3}$, which fitted into the series complete it as follows:

Vib. No.	256	288	320	341⅓	384	426⅔	480	512
Name	C	D	E	F	G	A	B	C'
Vib. Ratio	C	$\frac{9}{8}$C	$\frac{5}{4}$C	$\frac{4}{3}$C	$\frac{3}{2}$C	$\frac{5}{3}$C	$\frac{15}{8}$C	2C
Intervals		$\frac{9}{8}$	$\frac{10}{9}$	$\frac{16}{15}$	$\frac{9}{8}$	$\frac{10}{9}$	$\frac{9}{8}$	$\frac{16}{15}$

It is also found that three notes whose vibration-rates are as $10:12:15$ produce, when sounded together, an effect but little less agreeable than the major triad. Such a combination, together with the octave of the lowest, constitutes a *Minor Chord*. The minor diatonic scale is built up in a manner similar to the major scale.

448. Intervals of the Diatonic Scale. — An examination of the table in the last article shows that there are three different intervals in the diatonic scale, and that they succeed each other in a certain order. In preparing this series of notes, C was chosen as the key-note. Now if some other note, as D, be chosen as the key-note, and we multiply it by $\frac{9}{8}$ to obtain the next note, we have a note of $\frac{9}{8} \times 288 = 324$ vibrations, which differs slightly from E. Again, if we multiply 324 by $\frac{10}{9}$ we have 360, a note differing widely from any note of the series. If we multiply 360 by $\frac{16}{15}$, we have 384, which is note G. If we take other notes of the series as key-notes and proceed in this way, we shall see that if this order of intervals is to be maintained in every key, several new notes must be interpolated in the series. Again, if the minor chords are to be provided for, it will be necessary to introduce several additional notes. These new notes are designated as the *Flats* or *Sharps* of those between which they lie, and are obtained by raising (sharping) or lowering (flatting) these notes by the interval $\frac{25}{24}$.

449. Tempered Scales. — The interpolation of notes as described above increases the number of notes to such an extent — seventy-two to the octave — as to be unmanageable in the practice of music, particularly in the case of music designed for such instruments as the piano, where each note is fixed. It becomes necessary, then, to reduce the number of notes by changing the values of the intervals. This process is called *Temperament*. Of the several methods proposed by musicians, that of *Equal*

Temperament is the one generally adopted. It makes all the intervals from note to note equal, interpolates only one note in each whole tone of the diatonic scale, and thus reduces the number of notes in the octave to twelve. This scale of twelve notes is called the *Equally Tempered Scale*.

EXERCISES.

1. How many beats are produced in a second by two notes whose rates of vibration are respectively 291 and 388?
2. What is the wave-length for a note due to 538 vibrations per second at 16° C.?
3. Determine the pitch of a note from the following data:
Number of holes in siren disk 36; number of revolutions in 10 sec. when in unison 84.
4. A piano is usually tuned so that C is 261; find the pitch for each note of the gamut.
5. Why is there a rise in pitch of a locomotive whistle when the train is rapidly approaching, and a corresponding fall when moving away from you?
6. A tuning-fork is held over a tall glass jar, into which water is gradually poured until the maximum reënforcement of the sound is produced. This is found to be the case when the length of the column of air is 64.8 cm. What is the vibration-number of the fork, the temperature being 16° C.?

X. VIBRATIONS OF STRINGS.

450. Modes of Vibration. — Of the three ways in which strings may vibrate, the transverse vibration is the most important. When used for the production of sound, strings are fastened at their ends, stretched to the proper tension, and made to vibrate either by drawing a violin bow across them, striking them with a light hammer, or plucking them with the fingers. An examination of any stringed musical instrument, as a violin, will make it evident that by varying the tension, the length, or the mass of the wires or strings, tones of any desired pitch can be secured.

451. Laws of Strings. — In order to study the laws obtaining in the case of strings, an instrument called a *Sonometer* is used (Fig. 210). This may be nothing more than a wooden box about 150 cm. long, 15 cm. wide, and 10 cm. deep. The material should be pine, free from pitch, straight-grained, and about 1 cm. thick, except the top, which may be 5 mm. thick. The ends should be

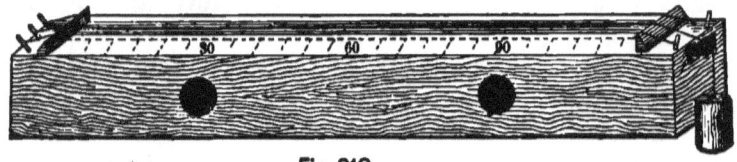

Fig. 210.

of hard wood, well seasoned, and have a thickness of about 25 mm. Near the ends are triangular pieces of hard wood, to serve as bridges across which to stretch the wires. The distance between them should be 120 cm. In one of the head-pieces are inserted two stout pins and in the other two piano posts for changing the tension of the wires. On the same end is a small pulley for one of the wires to draw over, the tension being measured by attached weights. Beneath the wires is a scale graduated to half centimetres.

Exp. — Stretch a wire on the sonometer and then tune it till the pitch of the sound is that given by the inner row of holes of the siren. Now shorten the wire one-half by means of a movable bridge; the sound will be found in unison, agreeing in pitch, with that given by the outer row of holes of the siren. Shorten it a fifth;

the sound will be in unison with that given by the next to the inner row. Shorten it a third; the sound will be in unison with that given by the next to the outer row. Hence,

The Law of Lengths. — *The tension and the diameter being constant, the vibration-number varies inversely as the length.*

Exp. — Stretch equally two wires of the same material having different diameters, and by means of a movable bridge shorten the heavier wire till it sounds in unison with the finer one. If we measure the diameters of the wires, we shall find that the ratio of these diameters is the inverse of that of their lengths, that is, $\frac{d}{D} = \frac{L}{l}$. But $\frac{L}{l} = \frac{N}{n}$, that is, L, the length of the shortened large wire, is to l, the length of the whole wire, as N, the number of vibrations made by the whole wire, to n, the number made by the short wire, or, what is the same thing, the number made by the small wire. Therefore, $\frac{d}{D} = \frac{N}{n}$. Hence,

The Law of Diameters. — *The tension and the length being constant, the vibration-number varies inversely as the diameter.*

Exp. — Stretch two wires of the same size with unequal but known weights. By means of the movable bridge placed under the wire of lower pitch, put the wires in unison. If we compare the length with the tension, it will be found that $\frac{L}{l} = \frac{\sqrt{T}}{\sqrt{t}}$, in which T and t represent the tension. But as shown in the last experiment $\frac{L}{l} = \frac{N}{n}$. Therefore, $\frac{N}{n} = \frac{\sqrt{T}}{\sqrt{t}}$. Hence,

The Law of Tensions. — *The length and the diameter being constant, the vibration-number varies as the square root of the tension.*

Exp. — Stretch equally a brass and a steel wire of the same diameter. By means of the movable bridge under the wire of lower pitch, put the wires in unison. If we compare lengths with densities of material we shall find that $\frac{L}{l} = \frac{\sqrt{s}}{\sqrt{S}}$, in which s and S represent densities. But $\frac{L}{l} = \frac{N}{n}$. Therefore, $\frac{N}{n} = \frac{\sqrt{s}}{\sqrt{S}}$. Hence,

The Law of Densities. — *The length, tension, and diameter being constant, the vibration-number varies inversely as the square root of the density.*

452. Applications. — In the piano, violin, harp, and other stringed instruments, the pitch of each string is determined partly by its length, partly by its tension, and partly by its size. The tuning is done by varying the tension.

XI. OVERTONES AND HARMONIC PARTIALS.

453. The Fundamental Tone of any vibrating body may be defined as the lowest tone that the body can give. It is produced when the body vibrates as a whole. In the

Fig. 211.

case of a string, its appearance is that of a single spindle (Fig. 211) whilst sounding its fundamental tone.

454. Nodes, Antinodes, etc. — **Exp.** — Attach a thread to the end of one of the prongs of a large tuning-fork. If such a fork cannot be had, it may be attached to the middle of one of the sonometer wires. Fasten the other end to a small rod held in a suitable clamp, so that the length and the tension may be varied at pleasure. Now set the fork in vibration, and vary both tension and the length of the string till it vibrates in a number of parts, giving it the appearance of a succession of spindles (Fig. 212).

SOUND. 293

The pulse started along *AB* by the fork is reflected at *B*, and, meeting new pulses on its return, produces points of no vibration at *N, N*, called *Nodes*. The points *V, V, V*, are those of greatest vibration, and are called *Antinodes*.

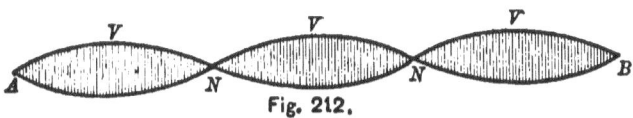
Fig. 212.

The portion of the wave from node to node, as *AN*, is called a *Ventral Segment*, and that from node to antinode, as *AV*, a *Semi-Ventral Segment*.

455. Overtones and Partials. — Exp. — Stretch a wire on the sonometer and set it in vibration by plucking or bowing it near one end. The tone heard most distinctly is its fundamental. Touch the wire lightly at its middle point. Instead of stopping the sound, a note an octave higher will be heard, showing that the wire is vibrating in two parts. If the wire be again plucked, both sounds can be heard together. Touching the wire one-third from the end brings out a tone an octave and a fifth higher, showing that the wire vibrates in three parts at the same time that it is vibrating as a whole. With a long string, it is possible to prove that a still further subdivision of a vibrating string takes place. In conducting such experiments, care must be exercised in selecting the point at which the string is plucked, for it is evident that there can be no node at that point.

The tones produced by sounding bodies vibrating in parts are called *Overtones* or *Partial Tones*. If the vibration-rate of an overtone is an exact multiple of the fundamental, it is called a *Harmonic Partial*. In strings the overtones may be harmonics, but in vibrating plates and membranes they are usually not so. The overtones are named *First, Second, Third*, etc., in the order of their vibration-rates as compared with that of the fundamental.

XII. VIBRATION OF AIR IN TUBES.

456. Gases as Sonorous Bodies. — It was seen in the use of the resonator that gases can be thrown into vibration when they are confined in tubes, and thus become sources of sound. Such a column of gas can be set in vibration in two ways: By a vibrating tongue, as in reed instruments, or by a stream of air striking against the edge of a lateral opening in the tube, as in the whistle, flute, etc.

457. Laws for Air-Columns. — **Exp.** — Select a glass tube of about 2 cm. diameter and 30 cm. long. To one end fit loosely a cork piston on the end of a stout wire. Flatten the end of a piece of brass tubing, and by means of it direct a stream of air across the mouth of the glass tube. If the position of the piston, as well as the force of the blast, be right, the tube will yield a pure tone. If we shorten or lengthen the air-column by moving the piston, the pitch of the tone will rise or fall accordingly. If we determine by trial the different lengths necessary to give the gamut, a comparison of them will give the continued ratio, $1 : \frac{8}{9} : \frac{4}{5} : \frac{3}{4} : \frac{2}{3} : \frac{3}{5} : \frac{8}{15} : \frac{1}{2}$, (447) showing that —

The pitch varies inversely as the length of the air-column.

Exp. — Prepare two glass or paper tubes, 20 and 10 cm. long, respectively, and about 2.5 cm. diameter. Hold the hand over one end of the shorter tube, and blow across the open end so as to produce its lowest pure tone. A comparison of this tone with that obtained by blowing across one end of the longer open tube will show that the pitch is the same. Since the open tube has twice the length of the closed tube when the pitch is the same, it follows that

The pitch of a closed tube is an octave below that of an open one of the same length.

458. State of Air in Sounding Tubes. — **Exp.** — Procure a small *Organ-Pipe*, made either of glass or with one glass side

(Fig. 213), and a small wire ring covered with thin paper. Sift a little fine sand over the paper membrane, and let it down by a thread into the tube supported vertically. Blow gently through the mouth-piece, producing the fundamental tone. At the same time move the ring up and down within the tube. The sand will be agitated the least when the membrane is near the middle of the tube, and the agitation increases as the membrane approaches either end.

Hence, when an open tube is sounding its fundamental, there is a node at the middle of the tube, and an antinode at each end.

If we close the end of the tube with a perforated cork, and again search for nodes, we shall find that the only node in a stopped tube sounding its fundamental is at the inner surface of the cork, and that there is an antinode at the other end of the tube.

459. Overtones. — **Exp.** — Blow a strong blast across the end of the longer tube used in Art. 457, and notice that notes of higher pitch than the fundamental are produced.

These tones of higher pitch are *Overtones*, caused by the air vibrating in parts or segments. In proof of this, insert a piston in the tube, and by means of it shorten the air-column till it gives as its fundamental the overtone previously obtained.

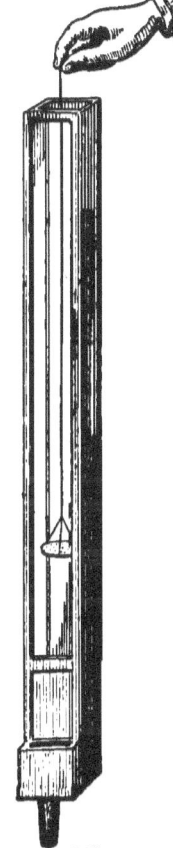

Fig. 213.

Since there can be no vibration of the air at the surface of the piston, it must mark a node, and since the tone is unchanged by the presence of the piston, there must have been a node at that point before the introduction of the piston.

460. Laws of Overtones. — An inspection of Fig. 214 will show that the following laws for overtones obtain:

I. *In open pipes the complete series of overtones is possible.*

II. *In closed pipes only those overtones are possible whose vibration-rates are 3, 5, 7, etc., times the fundamental.*

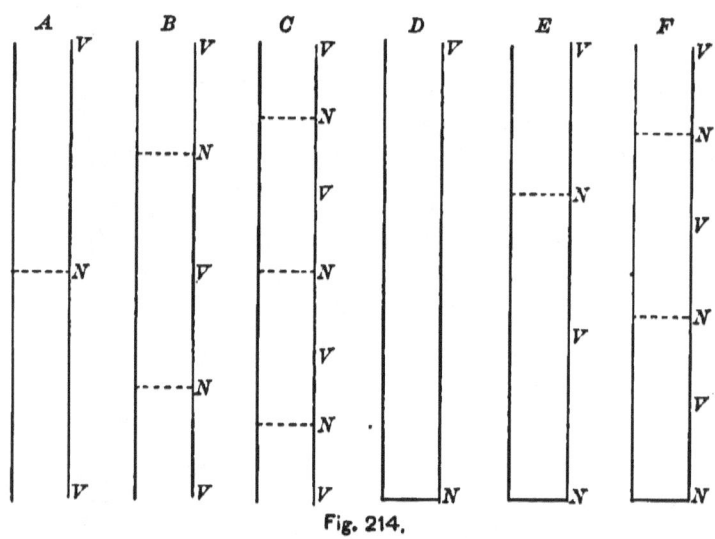

Fig. 214.

In Fig. 214 A represents the open pipe giving the fundamental; B represents the first overtone of A, and there is a node one-fourth from the end of the pipe; C represents the second overtone of A, and there is a node one-sixth from the end of the pipe. D represents the closed pipe giving the fundamental; E represents the first overtone of D, and there is a node one-third from the end of the pipe; F represents the second overtone of D, and there is a node one-fifth from the end of the pipe.

XIII. QUALITY OR TIMBRE.

461. Quality. — It has been shown that sonorous bodies can vibrate as a whole, and at the same time in parts, thereby giving to the sound a complex character, the sound-waves produced being the result of compounding the fundamental with its overtones. The form of the sound-wave will evidently depend on the relative phase and intensity of the overtones present, and this resultant wave-form will determine the *Quality* of the tone.

462. Applications. — Every one has doubtless noticed that sounds of the same pitch and intensity often differ so widely in character that there is no difficulty in distinguishing them; for example, the A of a piano is easily distinguished from the A of a flute, violin, or tuning-fork. The difference is due to the series of overtones present in each case. Voices differ for a like reason. Violins differ in sweetness of tone, for the reason that the sounding-boards of some bring out the overtones more perfectly than others. Voice culture consists in training and developing the vocal organs to the end that a better production of overtones may be secured and a greater richness be thereby imparted to the tones. Often in the reflection of sound by distant objects its character is greatly changed, owing to the suppression of overtones in the process of reflection.

463. Complex Character Proved. — Helmholtz was the first to demonstrate the correctness of the explanation given for quality in the preceding articles. He constructed a series of resonators, consisting of hollow metallic globes (Fig. 215) of different sizes, provided with two apertures of unequal diameters. The air within the globe has only one vibration-rate natural to it, and that depends on the cubic contents and the size of the orifices. The series of resonators used comprised one that reënforced a certain fundamental tone and one for each of its overtones. By placing these resonators successively to the ear, when a tone of the pitch of the fundamental is sounded, the presence of any of its overtones is made evident by the resonator responding. Proceeding in this way, Helmholtz was enabled to demonstrate that quality of sound is determined by the overtones present; that to give a tone richness and sweetness the first four or five overtones are essential.

Fig. 215.

Helmholtz also proved that any quality of sound may be built up by combining a tone with its overtones. To do this he employed a set of tuning-forks, kept in vibration by electro-magnets. The set consisted of ten forks, nine of which sounded the overtones of the tenth. Each fork was provided with a resonator to strengthen its sound. By means of a set of keys any particular resonator could be brought into action, and the sound of the fork made audible at a distance.

EXERCISES.

1. A string sounding C' is 18 inches long; how much must it be lengthened to sound the note D?

2. A string stretched by a weight of 4 lbs. sounds a certain note C; what weight will be needed to sound the F of the scale?

SOUND. 299

3. A stretched string 10 feet long is in unison with a tuning-fork marked 256; the string is shortened 4 ft.; how often will it now vibrate in a second?

4. A string is fastened at one end to a peg in a horizontal board, and the other end passes over a pulley and carries 16 lbs. The string gives the note C. What weight must be used instead of the 16 lbs. so that the string will give the octave below?

5. A wire stretched by a weight of 13 kilos. sounds a certain note. What must be the stretching weight to produce the major third?

6. Compute the lengths of open pipes necessary to give the tones of a major chord at a temperature of 16° C., calling C = 256.

XIV. HARMONY AND DISCORD.

464. Consonance and Dissonance. — Two musical sounds of the same vibration-rate are said to be in *Unison*, and their combination produces a smooth and pleasant tone. When two notes differ in pitch and their combination is agreeable to the ear, they are said to be *Consonant;* when disagreeable, *Dissonant*.

465. Cause of Dissonance. — Helmholtz proved that dissonance is due to beats (435), and is governed by the following rule:

If the number of beats per second between the fundamentals of the two tones, or between their overtones, or between the overtones of the one and the fundamental of the other, is between 10 and 70, dissonance will occur.

He found that maximum dissonance was caused by about 30 beats per second, and that consonance was admissible at 70, and all roughness disappeared at about 130.

To illustrate the above rule, let us apply it to the notes C and D, C and E, and C and G.

		C	D	C	E	C	G
Fundamental tone,		256	288	256	320	256	384
1st overtone,		512	576	512	640	512	768
2d	"	768	864	768	960	768	1152
3d	"	1024	1152	1024	1280	1024	1536
4th	"	1280	1440	1280	1600	1280	1920
5th	"	1536	1728	1536	1920	1536	2304
6th	"	1792	2016	1792	2240	1792	2688
7th	"	2048	2304	2048	2560	2048	3072

Comparing these tones, we find a difference of 32 between the fundamentals C and D, showing 32 beats. This will be strengthened by 64 between their first overtones, also by 64 between the fifth overtone of D and the sixth of C, and also by 32 between the sixth of D and the seventh of C. Hence, the tones C and D will yield maximum dissonance.

The fundamentals C and E give 64 beats, which is supported by 64 between the second of E and the third of C, and also by 64 between the fourth of E and the fifth of C. The result is a slight dissonance, or an imperfect consonance.

Between C and G there is no case of less than 128 beats; and we have in them no perceptible dissonance.

XV. VIBRATING RODS, PLATES, AND BELLS.

466. Vibration of Rods. — Rods of metal, of wood, and of glass may be made to vibrate either transversely or longitudinally. The former mode may be produced by fixing the rod and drawing a violin bow across the free end, or by striking it with a suitable hammer; the latter by clamping the rod at the middle and stroking it lengthwise with a cloth dusted over with powdered resin. The Jew's harp, music box, and gong of a clock, when made of coiled wire, are illustrations of vibrating rods clamped at one end. The tuning-fork may be regarded as an elastic bar free at the ends, and supported in the middle by a stem which is subject to all the motion of the middle of the ventral segment, giving it an up and down motion which is transmitted to the supporting sounding-board (Fig. 216). The overtones possible are of very high pitch and feeble intensity, and soon vanish, leaving the tone pure.

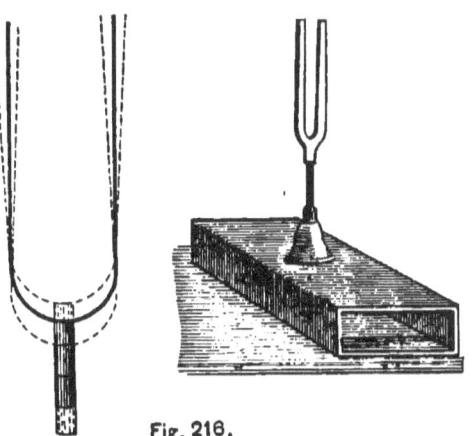

Fig. 216.

467. Vibration of Plates. — **Exp.** — Support a square brass plate as shown in Fig. 217. Scatter a little fine sand evenly over it, touch the plate at some point with the finger, while a violin-bow is drawn across the edge. The plate is thrown into vibration, the sand arranging itself in symmetrical figures whose complexity increases as the pitch becomes higher.

Fig. 217.

A study of these sand figures makes it clear that the

plate vibrates transversely in segments, the sand being thrown to places of least vibration which lie between parts having opposite motions. The arrangement of nodal lines is determined by the relative position of the point bowed to that pressed by the finger. This method of studying vibrating bodies was first employed by Chladni.

468. Vibration of Bells. — **Exp.** — Draw a violin-bow across the edge of a large bell or goblet half full of water. It will yield a musical sound, and at the same time the surface of the water will be covered with ripples proceeding from the several segments into which the vibrating body is divided.

The lowest tone is obtained when the bell divides into four segments, the pitch rising with the number of segments. Powdered sulphur sifted evenly on the water will render more conspicuous the position of the nodes.

XVI. GRAPHIC AND OPTICAL STUDY OF SOUND.

469. Graphic Methods of studying sound are of service in determining the vibration-rate of sounding bodies.

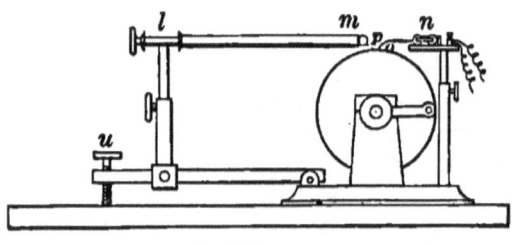

Fig. 218.

In one of the simplest a small style is attached to the vibrating body and traces its movements upon a piece of smoked paper or glass, which moves regularly beneath it. Generally, the paper is wrapped around a cylinder, mounted on an axis, one end of which has a screw-thread on it, so that when the cylinder turns it also moves axially (Fig. 218). If the vibrating body is a fork, by measuring the time and counting the number of sinuosities in the line traced on the cylinder, the rate of the fork is easily ascertained.

470. **The Phonograph** is an apparatus invented by Edison for reproducing sound. Its essential features are:

1. A mouth-piece into which the speaker talks. At the bottom of this is a thin elastic disk carrying a sharp graver on its under side.

2. A metallic cylinder mounted on a screw axis and arranged for regular rotation, either by a crank or an electric motor.

3. A hollow cylinder of wax, covering the metallic cylinder, in which the graver, moved by the sound-waves striking the elastic disk, cuts a shallow furrow varying in depth with the wave-form.

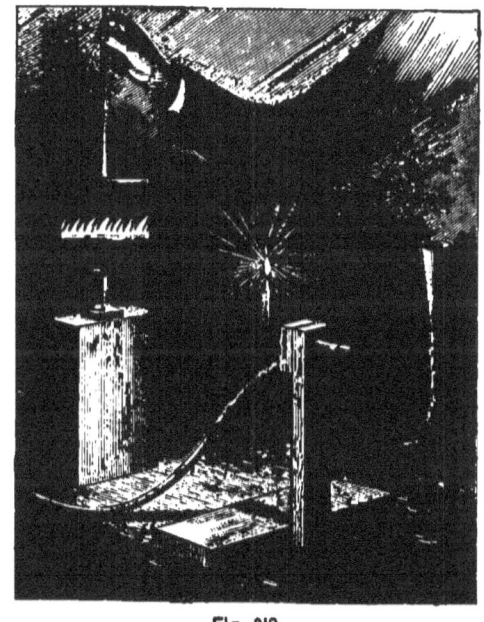

Fig. 219.

To reproduce the sound a smooth rounded style, attached to a thin disk, follows the furrow as the cylinder is turned, and the indented bottom imparts to the style and disk vibrations similar to those which caused the indentations.

471. Manometric Flames. — Exp. — Cut out of a board, 2.5 cm. thick, a strip 4 cm. wide and 25 cm. long. Near one end bore a hole 2.5 cm. in diameter, half way through it. Cut a second piece of the same width and 5 cm. long, and bore into it a similar hole. Screw these two pieces together so that the two holes face each other, separating them, however, by a membrane of gold-beater's skin, thus forming a box divided into two parts by this elastic partition. Fasten the long strip to a suitable board for a support (Fig. 219). Bore into the box three holes, two in one face and one in the other. Into the latter cement a glass tube for the purpose of attaching a short rubber tube leading to a funnel-shaped mouth-piece. In the other holes cement a jet-tube, and a straight tube for making connection with the gas supply.

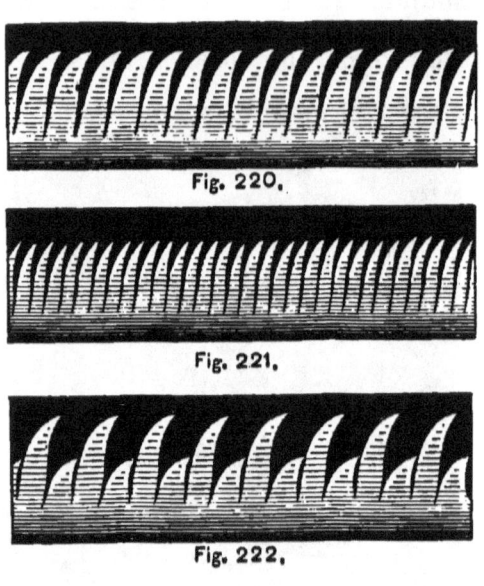

Fig. 220.

Fig. 221.

Fig. 222.

Admit gas to the box and light it as it escapes from the jet-tube. Rotate in front of it a small plane mirror mounted on a spindle. The image of the flame will be a smooth band of light. Now sound a heavy C-fork in front of the mouth-piece or produce there any pure tone; the appearance in the rotating mirror will be that of Fig. 220. A condensation entering the box acts on the membrane, compressing the gas, and thereby extending the flame; a rarefaction entering it produces an opposite effect. Hence, a serrated band is seen in the mirror. Now sound a C'-fork, a simple tone an octave higher than that first used; the appearance is that shown in Fig. 221, differing from the last in the teeth being half as wide. If we connect two mouth-pieces to

SOUND. 305

the box, using a T-pipe, and sound the C-fork in front of one, and C′-fork in front of the other, we obtain Fig. 222, the short tongues of which are due to C′, the octave of C. The same figure is obtained by singing the vowel sound *o* on the note B♭.

The experiment shows the possibility of analyzing sounds by the flame pictures they may be made to produce. The method is due to Koenig of Paris, and has the great advantage of being independent of the sense of hearing. If this box or manometric capsule, as it is called, be attached to a Helmholtz resonator, the flame will respond whenever any sound is produced that affects the resonator. With a complete set of resonators, each with its manometric capsule, a most efficient apparatus is provided for the analytical study of sounds.

CHAPTER VII.

LIGHT.

I. ITS NATURE.

472. Light is the immediate external cause of the sensation of sight. It is believed to be a periodic disturbance in a very subtile and highly elastic medium which is supposed to exist everywhere in space, even pervading the intermolecular spaces in matter. This medium is known as the *Ether*, and vibratory disturbances in it gives rise to all the phenomena of radiant energy. These disturbances are propagated through it as waves, not of compression and rarefaction, but more like those of the rope (400), the direction of vibration being transverse to that of propagation.

473. A Luminous Body is one that emits light. Such a body can send light to the eye, and hence is visible. Bodies which emit light of themselves are said to be *Self-Luminous*. The sun, stars, bodies heated to incandescence, and matter undergoing rapid combustion are familiar examples. When a luminous body emits the light it has received from some other body, it is said to be *Illuminated*. Most of the bodies around us belong to this class.

474. Media Classified. — In general, when light is incident on a body, a part of it is reflected, a part is transmitted, and the rest is absorbed. *Transparent* bodies allow light to pass through them with so little loss that objects can be easily distinguished through them; as glass, air, pure water. *Translucent* bodies transmit light so very imperfectly that objects cannot be seen distinctly through them; as horn, oiled paper, very thin sheets of metal or wood. *Opaque* bodies transmit no light, as blocks of wood or iron. This classification is one of degree; no sharp line of separation between these classes can be drawn.

No body is perfectly transparent. If several layers of glass are superposed, the distinctness of vision through them diminishes with the increase in the number of layers; stars which are invisible at the foot of a mountain are often visible at the top. (Why?) Neither is any body perfectly opaque, since all substances when cut into very thin sections are more or less translucent.

II. THE PROPAGATION OF LIGHT. ITS VELOCITY.

475. Rays. — If an opaque object is placed between the eye and a luminous point, it hides the point from view, showing that *light is propagated in straight lines.* Certain facts which will be considered in a subsequent article (510) make it necessary to modify this statement by adding the restriction, *in a homogeneous medium.*

476. Rays of Light are the lines along which light is transmitted. These lines must be radii of spheres, since light is propagated as spherical waves whose origins are the points composing the luminous body. When the source of light is at a great distance, the several rays comprised in a bundle of them are sensibly parallel, and constitute a *Beam of Light;* for example, in the case of light from the sun or stars, the distance is so great that the rays incident on any small object are said to be parallel. Rays of light from a point form a *Diverging Pencil,* and rays toward a point, a *Converging Pencil.*

477. Shadows. — When light is incident on an opaque object, it is excluded from the space behind the object, since it is propagated along straight lines. This space is called the *Shadow.*

478. Parts of a Shadow. — **Exp.** — Hold a ball between the flame of a lamp and a white screen. The section of the shadow on the screen is not equally dark in all its parts. Now make several pin-holes in the screen, having at least one in the darker part, one in the lighter, and one outside of both. On looking through these holes toward the light, no part of the flame can be seen through any hole in the darker portion. A part of the flame can be seen through any hole in the lighter portion, and the whole flame can be seen through any hole outside of the shadow.

Hence, the darker part of the shadow, called the *Umbra,* is due to the total exclusion of the light by the opaque object, and the lighter part, the *Penumbra,* to its partial exclusion.

479. Different Cases. — *First.* When the luminous object is a point. Let L (Fig. 223) be a luminous point, and AB an opaque body. Then the shadow must be bounded by the cone of rays,

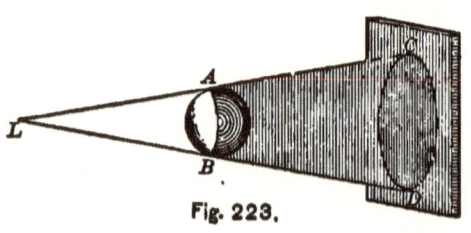

Fig. 223.

ALB, tangent to the object. In this case it is evident that the shadow has but one part, the umbra, and that its cross-section is greater as the distance from the object increases.

Second. When the luminous object has magnitude and is smaller than the opaque body. Let *LL* (Fig. 224) be a luminous object, and *AB* an opaque one. Since each point of the luminous object acts as a radiant point, the cone of light from each point and tangent to the opaque body forms a shadow.

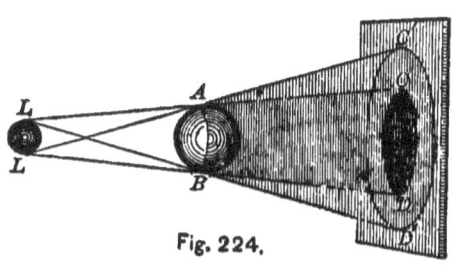

Fig. 224.

The space *ABDC* behind the opaque body and common to all the cones, evidently receives no light from *LL*; the parts between *AC* and *AC'*, and between *BD* and *BD'*, receive some light from the object, the amount increasing as *AC'* and *BD'* are approached. The cross-section of both the umbra and the penumbra increases as the distance from the object increases.

Third. When the luminous object is equal in size to the opaque one.

[Let the student draw the figure and discuss.]

Fourth. When the luminous object is larger than the opaque one.

[Let the student draw the figure and discuss.]

480. Images by Small Apertures. — **Exp.** — Support in vertical parallel planes two small sheets of cardboard, *A* and *B* (Fig. 225). In the centre of *A* cut a round hole about 2 mm. in diameter, and place in front of it, at a distance of a few centimetres, a lighted candle *CD*. There is formed on the other sheet an inverted image of the flame. If a triangular or a square hole is used, an equally distinct image of the candle is formed, showing that the image is independent of the shape of the aperture.

Now if the aperture is gradually enlarged, the image loses in distinctness of outline, gains in brightness, and gradually assumes the shape of the aperture.

Hence, the definition and the brightness of the image are independent of the shape of the aperture, but are governed by its size.

481. How Formed. — Each point of the object is the vertex of a cone of rays passing through the aperture and

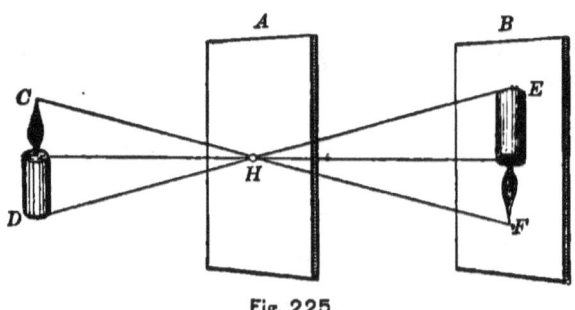

Fig. 225.

forming an image of it on the screen. These images will be symmetrically placed with reference to the points emitting the light, and consequently will build up a figure of the same form as the luminous object. Now it is evident that these numerous pictures of the aperture overlap in forming a picture of the object, the number superposed at any one place determining the brightness of that place. The edges of the picture will, therefore, be less bright than the other portions, and these differences will be more noticeable the larger the opening. In the case of a very large opening, the overlapping of the images of the aperture destroys all resemblance of the image to the object, the resulting image being that of the aperture.

482. Illustrations. — When the sun shines through the small chinks in the foliage of a tree, there may be seen on the ground a number of spots of light either round or oval. These are images of the sun. During partial eclipses of the sun such figures assume a crescent shape.

In the photographer's camera, and in the eye, we have further illustrations of the formation of images by small apertures. In these cases, as will be seen in subsequent articles, the definition is improved by the aid of a lens (521).

483. Velocity of Light. — Until 1675 it was believed that light travelled instantaneously. In that year, Roemer, a Danish astronomer, was engaged in studying the eclipses of Jupiter's moons, a phenomenon which takes place when Jupiter is between the sun and the eclipsed moon (Fig. 226). The moon nearest the planet is eclipsed at each revolution. Roemer noticed that if he determined by observation the interval between two successive eclipses as seen from the earth at E, and from this predicted the times of all the eclipses during the passage of the earth from E to E',

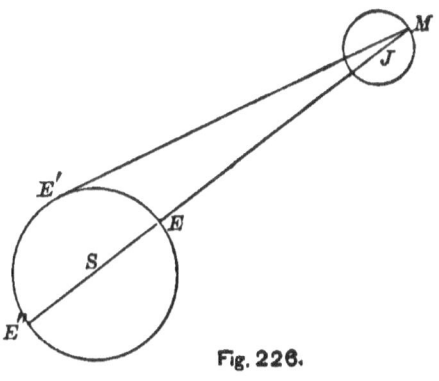

Fig. 226.

the eclipses fell constantly more and more behind, till at E'' the delay was about 16.5 minutes. If, on the other hand, the eclipses were predicted from the time when the earth was opposite Jupiter, he found that they occurred before the computed time. Roemer's explanation of the whole phenomenon is, that when the earth is at E'', the light received from the sun by way of the planet travels the diameter of the earth's orbit farther than when the earth is at E, and that the 16.5 minutes are consumed in traversing this increased distance. This assigns to light a velocity of about 186,000 miles per second. In 1849, Fizeau, by an ingenious device which enabled him to measure extremely short intervals of time, determined the time required for light to travel over a known distance on the earth's surface. Foucault, in 1850, showed not only that light occupies a measurable time in travelling any given distance, but also that the velocity varies with the medium. These experiments have been repeated by other investigators, and the results, as summed up by Prof. William Harkness, show that the velocity of light in air is about 186,337 miles (299,877.64 kilom.) per second.

III. PHOTOMETRY.

484. The Intensity of illumination is the quantity of light received on a unit of surface. Every-day experience shows that it varies not only with the nature of the source, but also with the distance at which the source is placed.

485. Law of Intensity. — **Exp.** — Cut from cardboard three squares, 4 cm., 8 cm., and 12 cm., respectively, mounting them on wire supports. The centres of these screens should be at the same distance above the table as the source of light. Use a lamp giving a small flat flame, and set it with the flame turned edgewise to the surface of the largest screen, and distant about one metre. Between the light and this screen, place the medium-sized screen so that it exactly cuts off the light from the lateral edges of the largest. If the source were a point, the light would now be cut off from the whole screen. In like manner place the smallest screen with reference to the intermediate. If these screens are located with care, it will be found that their distances from the light are as 1 : 2 : 3. Now, as each screen exactly cuts off the light from the one next to it in the series, it follows that each screen, if the light were not interrupted, would receive equal amounts of light from the source. The surfaces of these screens are as 1 : 4 : 9, and hence the amount of light per unit of surface must be inversely as 1 : 4 : 9; but 1, 4, and 9 are the squares of 1, 2, and 3 respectively.

Therefore, *the intensity of the illumination varies inversely as the square of the distance from the source of light.*

486. **A Photometer** is an instrument for comparing the intensity of one light with that of another assumed

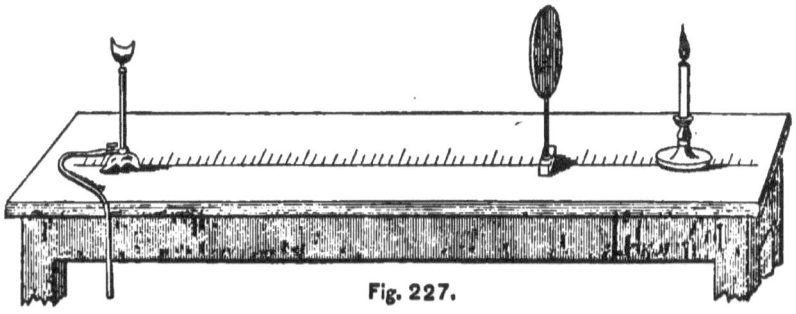

Fig. 227.

as a standard. The principle applied is that the ratio of the intensities of two lights equals the square of the ratio of the distances at which they give equal illuminations. The standard in general use is the light emitted by a sperm candle of the size known as "sixes" when burning 120 grains per hour. The illuminating power of a light is expressed by stating the number of times it is greater than the standard candle.

487. The Bunsen Photometer. — Exp. — Drop some melted paraffin on a piece of unglazed white paper. When cold remove all the excess with a knife. Heat the spot thoroughly with a hot flat-iron, in order to get the paraffin into the paper. The spot should have a diameter of about 3 cm. Mount the sheet of paper in a suitable frame, and support it in a vertical plane. Place the standard candle at one end of a long table, the light to be measured at the other, and the photometer disk between them, arranging the lights at the same height as the grease-spot and in a straight line with it (Fig. 227). By trial find a position for the candle where the grease spot is either invisible or is least conspicuous to one viewing it from a position in line with the three objects. When this occurs the disk is equally illuminated on the two faces. The square of the ratio of the distance of the light from the disk to that of the candle is the candle-power of the light. If, for example, the distances of a lamp and candle were 8 ft. and 2 ft. respectively, the candle-power of the lamp is $\left(\frac{8}{2}\right)^2 = 16$; that is, the lamp is equivalent to 16 standard candles in illuminating power.

EXERCISES.

1. In what respects does light differ from sound?
2. What must be the relative positions of the sun, earth, and moon, in order that there may be an eclipse of the sun?
3. What principle is recognized in aiming a rifle?
4. Why is the image of an object formed by a small aperture inverted?
5. Oculists often introduce atropine into the eye to enlarge the pupil. Why is one then unable to see objects distinctly?
6. If the distance of an object from the source of light is doubled, how is the size of the shadow affected? Show by figure.

IV. REFLECTION OF LIGHT.

488. Regular Reflection. — When a beam of light falls on a polished surface AC (Fig. 228), the greater part of it is reflected in a definite direction. The angle that an incident ray makes with the normal PB to the reflecting surface at the point of incidence is called the *Angle of Incidence*, as IBP; and the angle between the reflected ray and this normal is the *Angle of Reflection*, as RBP.

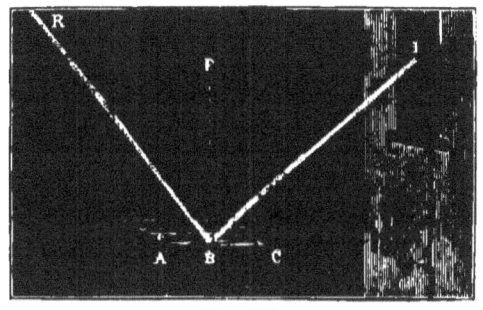

Fig. 228.

489. Law of Reflection. — Exp. — Cut from a board a piece somewhat greater than a semi-circle, giving it a radius of one foot. Pivot at the centre two arms, one having a candle or small lamp at its free end, and the other a screen of oiled paper (Fig. 229). At the centre of the circle mount a piece of mirror with its reflecting surface perpendicular to the radius that bisects the curved edge. Near the candle mount a convex lens (521). Between the lens and candle support a stout wire. If the lens and wire are given suitable relative positions an image or picture of the wire will be found across the centre of the paper screen whenever the arms make equal angles with the normal to the mirror.

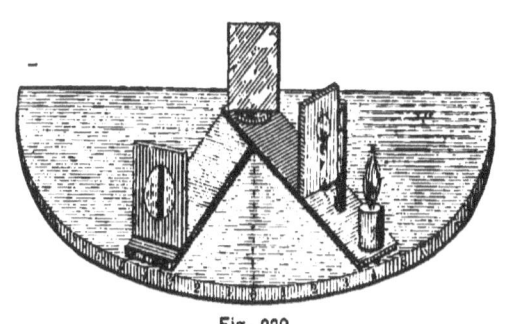

Fig. 229.

Hence, *The angle of reflection is equal to the angle of incidence, and the two angles lie in the same plane.*

490. Diffused Reflection. — Exp. — Fill a large glass jar with smoke. Cover the mouth with a piece of cardboard, through which is a hole about 1 cm. in diameter. With a small hand-glass reflect a beam of sunlight into the jar through the aperture in the cover. The whole interior of the jar will become illuminated.

The small particles of smoke floating in the jar furnish a great many surfaces. The light incident on them is reflected in as many different directions, the result being seen in the diffusion of the beam.

All reflecting surfaces, to a greater or less extent, scatter light in the same way as these smoke particles, on account of the irregularities of their surfaces. Fig. 230

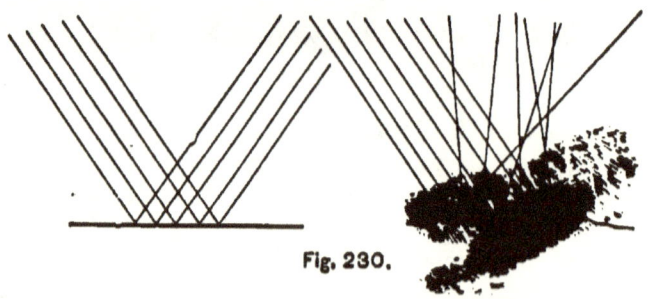

Fig. 230.

illustrates the difference between a perfectly smooth reflector and reflectors as they actually are, more or less irregular according to the degree of polish.

491. Benefits from Diffused Reflection. — It is by diffused reflection that objects become visible to us. Perfect reflectors would be invisible. The trees, the ground, the grass, and particles floating in the air reflect in every direction the light from the sun, and thus fill the space about us with light. Aeronauts tell us that as they reach very high altitudes the sky grows black, owing to the absence of floating particles to diffuse the light.

492. A Mirror is any smooth surface. A *Plane Mirror* is one whose reflecting surface is plane. A *Spherical Mirror* is one whose reflecting surface is a portion of the surface of a sphere.

493. Images. — If an object be placed in front of a mirror, an apparent object formed by the reflected light will be seen. This is called an *Image*. When the reflected rays come to the eye apparently but not actually from the place assigned to the image, the image is *Virtual;* but when the reflected rays unite to form an image, and the light comes from the image to the eye, it is *Real*. Such an image can be received on a screen.

494. Images in Plane Mirrors. — *First.* Let A be a luminous point in front of a plane mirror MN (Fig. 231). Any ray AB incident on the mirror is reflected in the direction BD, making the angle of reflection FBD equal to that of incidence FBA. In like manner, a second ray, as AC, is reflected along CE. If BD and CE are produced, they meet at A'. Join A and A'. Then, by geometry,[1] AA' is perpen-

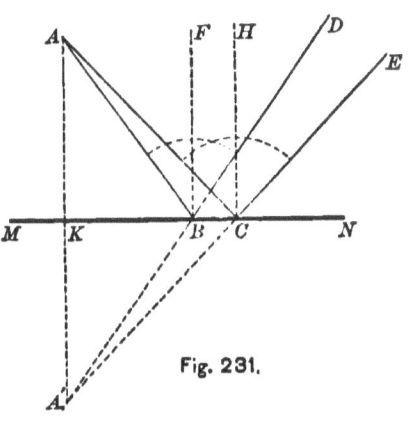

Fig. 231.

---

[1] The angle $ABK = DBN = A'BK$. ∴ $ABN = A'BN$. The angle $ACK = ECN = A'CK$. Hence, the triangles ABC and $A'BC$ are equal, for they have two angles, and the included side of the one equal to the corresponding parts of the other, and therefore $AB = A'B$. Comparing the triangles AKB and $A'KB$, we have the angles at K equal, and hence right angles, and $AK = A'K$, since the triangles have two sides, and the included angle of the one equal to the corresponding parts of the other.

dicular to MN, and A' is as far back of MN as A is in front. Since AB and AC are any two rays diverging from A, and incident on the mirror, it follows that *all* rays from A, *incident* on MN, must be reflected from MN as if they came from a point as far back of MN as A is in front. Hence, A' will appear to the eye like A; that is, an image of A is seen at A'. Therefore, *the image of a point in a plane mirror is virtual, and is as far back of the mirror as the point is in front;* it may be found by drawing from the point a perpendicular to the mirror, and producing it till its length is doubled.

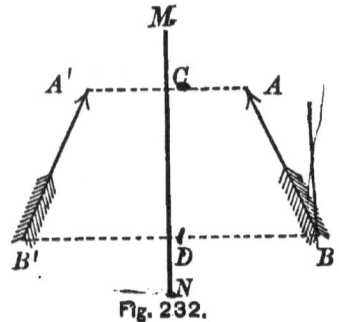
Fig. 232.

Second. Since the image of an object is an assemblage of the images of its points, it follows that the image of an object can be located by locating the images of its points. Let AB (Fig. 232) represent an object in front of the mirror MN. Draw the perpendiculars AA' and BB', making $A'C = AC$, and $B'D = BD$. Join A' and B'. Then $A'B'$ is the image of AB, and is evidently virtual, erect, and of the same size as the object.

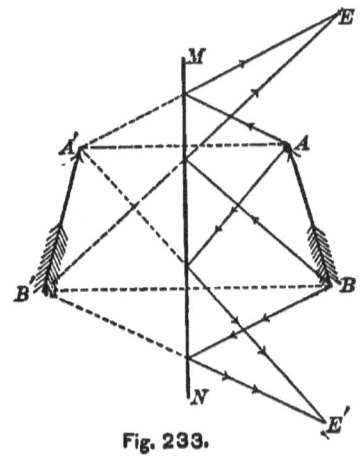
Fig. 233.

495. Path of the Rays to the Eye. — The rays which form the image for one observer are not those which form it for another. Let E and E' (Fig. 233) represent two different observers. To find the path of the rays which enter the eye at E, locate the image $A'B'$ as directed in the last article; then draw lines from A' and B' to E. The intersections of these lines with MN are the points of incidence of the rays from A and B which are reflected to E. In like manner, the rays which reach E' are determined.

496. Experimental Proof of the Position of the Image. — **Exp.** — Support a pane of window-glass in a vertical plane. On one side of it place a lighted candle, and on the other a tumbler of water, each at the same distance from the glass, and in a line perpendicular to it. An image of the candle will be seen in the tumbler of water, showing that the image is as far back of the mirror as the object is in front.

The experiment also explains how many optical illusions, such as Pepper's ghost, are produced. A large sheet of unsilvered glass, with its edges hidden from view by curtains, is so placed that the audience have to look through it to see the actors on the stage. Other actors, strongly illuminated and out of view by the audience, are seen by reflection in the glass and appear as ghosts on the stage. The magic cabinet and the disembodied head are also illusions produced by the aid of mirrors.

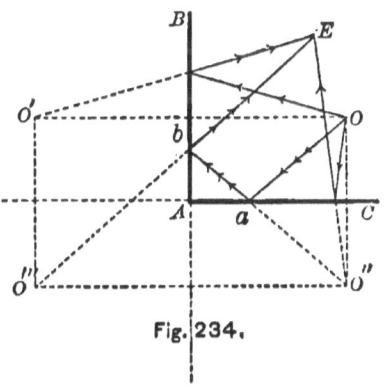

Fig. 234.

497. Multiple Reflection. — **Exp.** — Place two mirrors at right angles with their reflecting surfaces toward each other (Fig.

320 ELEMENTS OF PHYSICS.

234). Hold a lighted candle between them. Three images of it will be seen. On varying the angle, we find that the number of images varies, increasing as the angle decreases, becoming very great when the mirrors are parallel.

' The effect is due to the repeated reflection from one mirror to the other and finally to the eye. The images O' and O'' are formed by single reflections from the mirrors AB and AC separately (494). But the image O''' is due to a second reflection, the light from O being incident on AC, from which it is reflected to the mirror AB, which in turn sends it to E. The point b is found by drawing the line $O'''\,E$, and a by the line $O''\,b$. O''' is to be regarded as the image of O'' in AB and the image of O' in AC; and since it is situated *back* of the plane of both mirrors, no images of it will be possible.

The number of images is governed by the angle between the mirrors. When it is an aliquot part of 360°, the number is one less than the number of times it is contained in 360° — provided this number is even.

498. Illustrations. — The double image of a bright star or a candle seen in a hand-glass is an example of multiple reflection, the front surface of the mirror and the metallic surface serving as parallel surfaces. Theoretically the number of images is infinite; only two, however, are seen, on account of their faintness. The *Kaleidoscope*, a toy invented by Sir David Brewster, is an interesting application of the same principle. It consists of a tube containing three mirrors extending its entire length, the angle between any two of them being 60°. One end of the tube is closed by ground glass, and the other by a cap provided with an aperture. Pieces of colored glass are placed loosely between the ground glass and a plate of clear glass parallel to it. On looking through the aperture at any source of light, symmetrically arranged images of these pieces of glass are seen, forming figures of great beauty, which vary in pattern with every change in the positions of the objects.

499. Spherical Mirrors are either *Concave* or *Convex*, according as the inner or the outer surface of the spherical shell is the reflecting one. The centre of the sphere is the *Centre of Curvature* of the mirror. Any straight line drawn through the centre is an *Axis*, and is perpendicular to the reflecting surface. The middle point of the mirror is its *Vertex*, and that axis which passes through it is the *Principal Axis;* other axes are known as *Secondary Axes*. The *Aperture* of the mirror is measured by the angle formed at the centre by the axes which pass through the extreme points of the circumference. Let *MN* (Fig. 235) be a curved mirror.[1] Then *C* is the *Centre of Curvature*, *BA* is the *Principal Axis*, *DE* and *HK* are *Secondary Axes*, and *MCN* measures the *Aperture*.

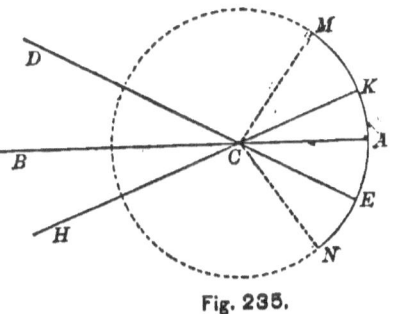

Fig. 235.

500. A Focus is a point towards which rays of light converge, or from which they diverge. When rays parallel to the principal axis are incident on the mirror, their paths after reflection are found to have a common point called the *Principal Focus*, and its distance from the mir-

---
[1] It should be noticed that the figures of mirrors employed in this chapter are sections by plane normal to the surface.

ror is the *Principal Focal Length*. The foci of curved mirrors may be either *Real* or *Virtual; Real* if the reflected rays meet in front of the mirror; *Virtual* if they seem to diverge from a point behind the mirror.

501. Foci of Concave Mirrors. — A clear understanding of this subject may be reached through the consideration of the following cases:

First. When the rays are parallel to the principal axis, that is, the radiant point is at an infinite distance.

Let MN (Fig. 236) be a concave mirror whose centre of curvature is C, vertex A, and principal axis BA. Let ED be parallel to BA. Then CD is the perpendicular at D, and CDF, the angle of reflection, must equal EDC, the angle of incidence. The ray BA will be reflected back on itself, being perpendicular to the mirror. Hence, the rays ED and BA have a common point F. If the point D is close to A, the point F will be found on measurement to be midway between C and A. On drawing other rays parallel to CA, and determining their course after reflection, they will be found to pass nearly through F. Hence, *the Principal Focus is real and is halfway between the centre of curvature and the vertex of the mirror.*

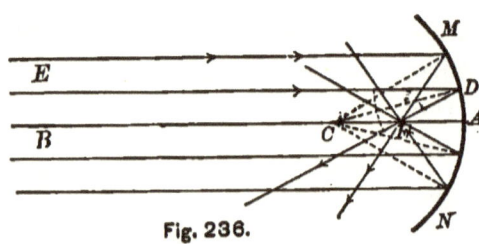

Fig. 236.

Second. When the rays diverge from a point beyond the centre of curvature.

Let the rays BA and BD diverge from B and be incident on the mirror MN (Fig. 237). BA will be reflected back on itself (why?), and BD will be reflected as DH, making with the perpendicular CD the angle CDH equal to the angle of incidence, BDC. The point B' common to these rays is the focus, *being real and situated between the centre of curvature and the principal focus.*

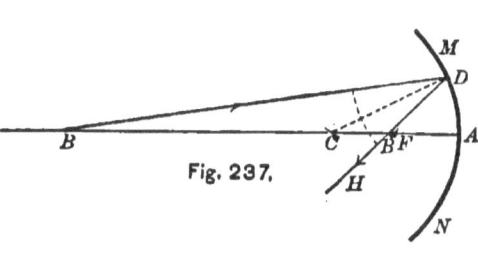

Fig. 237.

Third. When the rays diverge from the centre of curvature.

Since the rays are perpendicular to the mirror, they will be reflected back on themselves, making *the focus coincident with the radiant point.*

Fourth. When the rays diverge from a point between the principal focus and the centre of curvature.

This is evidently the converse of the second case, for the rays $B'A$ and $B'D$ (Fig. 237) will focus at B. Points having these properties are known as *Conjugate Foci.*

Fifth. When the rays diverge from the principal focus.

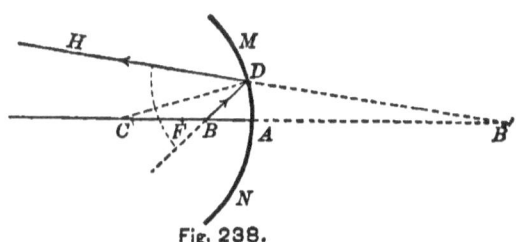

Fig. 238.

This is the converse of the first case, the rays leaving the mirror parallel to the principal axis, *giving no focus.*

Sixth. When the rays diverge from a point between the principal focus and the mirror.

Let the rays BD and BA diverge from B and be incident on the mirror MN (Fig. 238). BA will be reflected back on itself, and *BD* will be reflected as *DH*. (Why?) Then *B'*, back of the mirror, is the point common to these reflected rays. Hence, *the focus is virtual.*

502. Foci of Convex Mirrors. — Let *MN* (Fig. 239) be a convex mirror whose centre of curvature is *C* and vertex *A*.

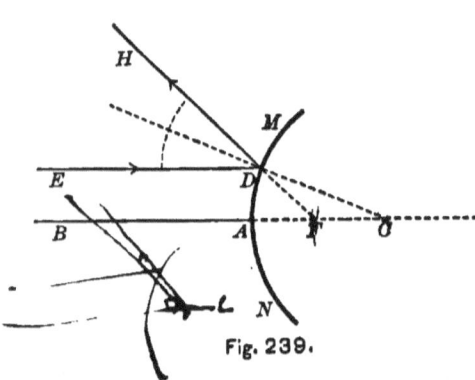

Fig. 239.

The ray *BA* will be reflected back on itself, *ED* will be reflected along *DH*, the incident and reflected rays making equal angles with the perpendicular *CD*. *F* is the point common to these rays, and it is half-way between *A* and *C*. *The principal focus is virtual, and is mid-way between the centre of curvature and the vertex.*

When the rays diverge from a point, as *B*, the focus will be virtual and will be situated between the principal focus and the mirror (Fig. 240). It will also be found that the general effect is not changed if the degree of divergence of the rays is changed. Hence, *the focus of a diverging pencil incident on a convex mirror is always virtual.*

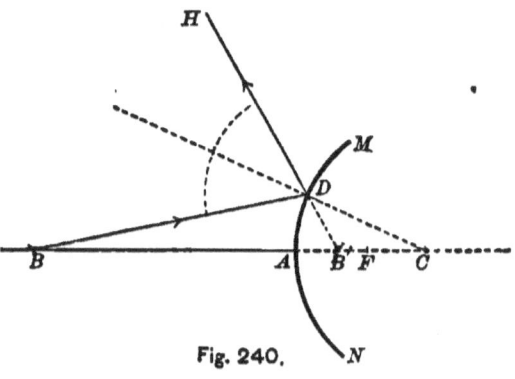

Fig. 240.

503. Formation of Images. — The image of an object

formed by a curved mirror may be determined by finding the images of its points. The work of constructing a figure is greatly simplified by observing the following rule:

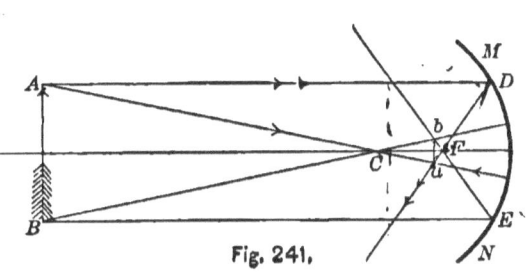
Fig. 241.

Draw secondary axes (499) through the extremities of the object. Rays along these lines will retrace their own path after reflection. Also draw through these same points rays parallel to the principal axis. These will be reflected through the principal focus. The intersection of two reflected rays which come from the same point of the object locates the image of that point.

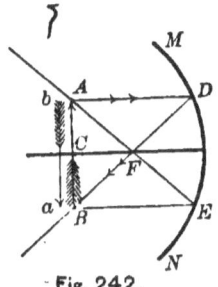
Fig. 242.

For example, in Fig. 241 the rays AD and AC are reflected through a, and BE and BC are reflected through b. Hence ab is the image of AB.

An examination of Figs. 241–246 will establish the following facts:

First. When the object is at an infinite distance the rays are parallel, and converge to the principal focus, making the image a point.

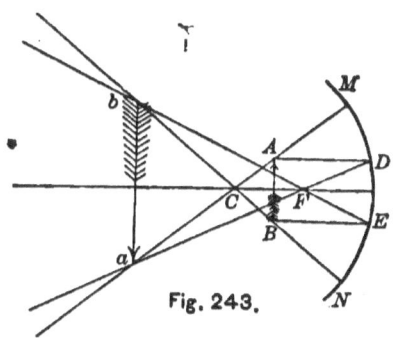
Fig. 243.

Second. When the object is at a finite distance, and beyond the centre of curvature, the image is real, inverted, smaller than the object, and situated between the centre of curvature and the principal focus.

Third. When the object is at the centre of curvature the image is real, inverted, of the same size as the object, and situated at the centre of the curvature.

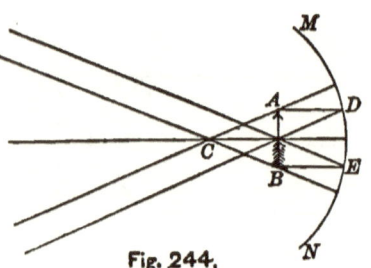

Fig. 244.

Fourth. When the object is between the centre of curvature and the focus, the image is real, inverted, larger than the object, and situated beyond the centre of curvature. (Compare this with the second case.)

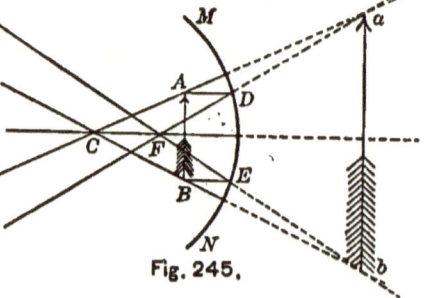

Fig. 245.

Fifth. When the object is at the focus the rays are reflected parallel, no image being formed. (See the first case.)

Sixth. When the object is between the focus and the mirror the image is erect, virtual, and larger than the object.

Seventh. The images formed by convex mirrors are always virtual, erect, and smaller than the object.

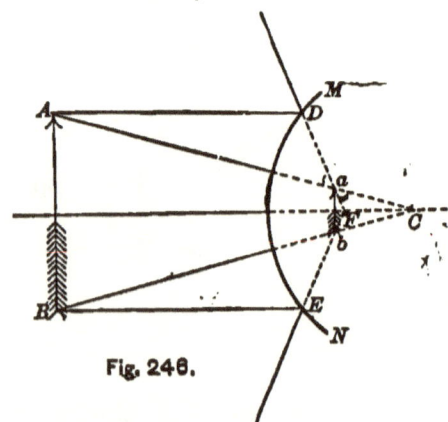

Fig. 246.

504. Experimental Proof. — In a dark room support vertically on a long table a spherical mirror of known focal length, and place in front of it a lighted lamp. When the image is virtual it can be seen by looking in the mirror; when real it can be received on a sheet of white cardboard held in front of the mirror, the exact place being found by trial. It will be necessary to place the lamp on a secondary axis, otherwise the screen will cut off the light from the object and prevent the formation of any image.

505. Focal Length Determined. — Exp. — In a sheet of cardboard cut a round hole about 1 cm. in diameter. Across the hole cement two threads at right angles. Place this screen in front of a concave mirror. Illuminate the hole and cross-threads by placing a lamp on the opposite side from the mirror. Now move the mirror till the image of the aperture formed on the screen is well defined, of the same size as the aperture, and situated close to the opening. Then half the distance of the screen from the mirror is the focal length. (Why?)

506. Spherical Aberration. — Exp. — Project an image of a lighted lamp on a screen with a concave mirror. The outer edge is indistinct; that is, the image is not sharply defined. Now cover up the reflecting surface, exposing only the central portion. The image will be less bright (why?), but the definition will be sharper.

This indistinctness is due to *Spherical Aberration*, all the rays from any point of the object not being focused by the mirror at the same point. This is made evident by tracing the course of the several rays of a beam of light incident on a mirror of wide aperture (Fig. 247). To prevent spherical aberration, the aperture of the mirror must be small, not exceeding 10°. Parabolic mirrors are free from this defect, and hence are used as

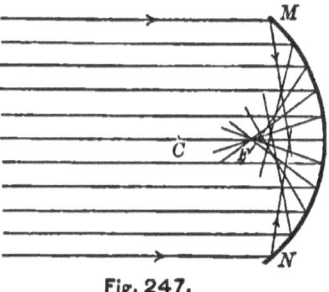
Fig. 247.

reflectors in light-houses and in the headlights of locomotives.

507. The Caustic. — **Exp.** — Bend a strip of bright tin into as true a semicircle as possible. Place it on a sheet of white paper with its concave surface toward the source of light (Fig. 248). The reflected light focuses along a curved line, called a *Caustic.*

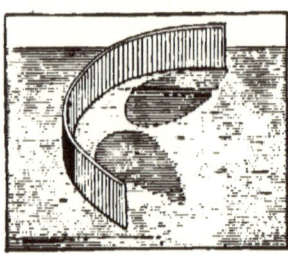

Fig. 248.

The caustic may be seen very distinctly by allowing sunlight to fall on a new tin basin partly full of milk, or on a plain gold ring resting on a white surface.

EXERCISES.

1. Show by a figure how it would be possible to see around an opaque object by the aid of mirrors.
2. Why are the images seen in multiple reflection of unequal brightness?
3. Why are the images of trees and of other objects seen in the water always inverted?
4. What should be the position of the reflector of a wall lamp in order to secure the best illumination of the room?
5. Place two plane mirrors about the grease spot used in the Bunsen photometer so that the observer standing in the plane of the photometer disk can see both faces by looking in the mirrors, and thus compare them for equality of illumination.
6. A plane mirror is inclined to a horizontal plane at an angle of 45°. Find the position of the image of a vertical object in such a mirror.
7. When a plane mirror is turned about an axis in its own plane, show by a diagram that the change in angle between the incident and the reflected ray is double the angle through which the mirror turns.

V. REFRACTION OF LIGHT.

508. Refraction. — Exp. — Place a rectangular tank, provided with a glass face, so that the light from a candle passing over the upper edge of one end just illuminates the whole of the opposite end (Fig. 249). The bottom of the tank lies wholly in the shadow cast by the end. Now fill the tank with water. The shadow no longer covers the whole bottom, since the rays are bent at an angle at the edge of the tank.

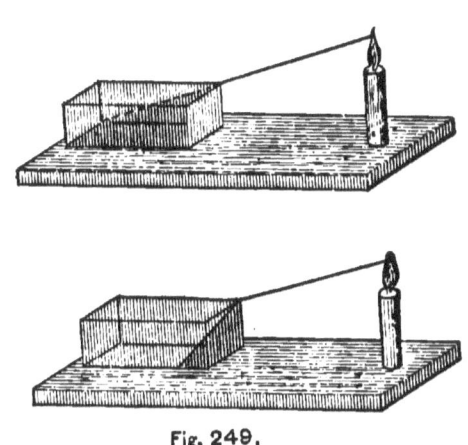

Fig. 249.

This bending of a ray when it passes from one medium to another is called *Refraction*.

509. Terms Defined. — Let *BA* (Fig. 250) represent a ray of light in air incident obliquely at *A* upon the surface *MN* of another medium, as water. Draw the normal *DE* to the refracting surface. The angle between the incident ray and the normal to the surface at the point of incidence is the *Angle of Incidence*, as *BAD*; and that between the refracted ray and the normal is the *Angle of*

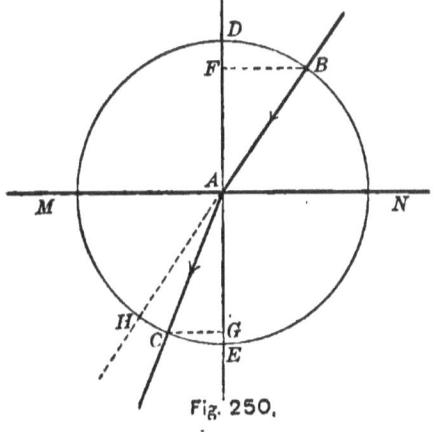

Fig. 250.

Refraction, as *CAE*. Produce *BA* to *H*. The angle *CAH* measures the amount by which the ray is bent out of its original course, and is known as the *Angle of Deviation*. With *A* as a centre, and a radius of one unit, describe a circle. Draw *BF* and *CG*, perpendicular to *DE*. Then *BF* and *CG* are the sines of the angles *BAD* and *CAE* respectively. The ratio of the sine of the angle of incidence to that of refraction, that is, $\frac{BF}{CG}$, is the *Index of Refraction*.

510. Laws of Refraction. — Exp. — Cover one face of a large rectangular glass battery-jar with paper, and out of the centre of the covering cut as large a circle as possible (Fig. 251). Across this circular opening draw a horizontal and a vertical diameter. Graduate the margin of the circle to degrees, the extremities of the vertical diameters to be marked 0°, and those of the horizontal, 90°. Provide a strip of cardboard considerably wider and longer than the top of the jar, and cut in it, crosswise, a slit 2 mm. wide and 5 cm. long. Fill the jar with water exactly to the horizontal diameter of the circle. Place the cardboard containing the slit on the top of the jar. With a plane mirror, reflect a beam of sunlight through the slit at such an angle as to be incident on the liquid exactly back of the centre of the circle on the face. Read on the circular scale the values of the angles of incidence and refraction and compute the ratio of their sines.[1] By changing the position of the slit and the mirror, other angles can be compared.

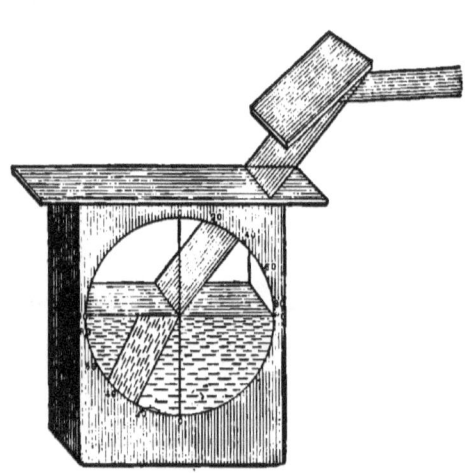

Fig. 251.

---

[1] The value of these sines is obtained by the use of trigonometrical tables.

LIGHT.

By proceeding as above it can be shown that light is refracted in accordance with the following laws:

I. *When a ray of light passes obliquely from a less refractive to a more refractive medium, it is bent toward the normal; when it passes in a reverse direction it is bent from the normal.*

II. *Whatever the obliquity of the incident ray, the ratio of the sine of the angle of incidence to the sine of the angle of refraction is constant for the same two media.*

III. *The planes of the angles of incidence and refraction coincide.*

511. Indices of Refraction. — By the *Absolute* index of refraction is meant the ratio of the sine of the angle of incidence to the sine of the angle of refraction when the ray passes from a vacuum into the substance. The *Relative* index of refraction is the index for light passing from one medium into another. Since the absolute index of air is very small, for most purposes it may be neglected. The relative index of refraction from one medium to another is equal to the inverse ratio of their absolute indices. The larger the index of refraction, the greater is said to be the optical *density* of the substance.

The following table gives the indices of a few substances:

Air at 0° C. and 760 mm.,	1.000294	Flint glass . .	1.576 to 1.642
Water	1.336	Crown glass .	1.531 to 1.563
Alcohol	1.372	Diamond . . .	2.44 to 2.755
Carbon bisulphide . .	1.678		

For the purposes of this book, water will be taken as $\frac{4}{3}$, crown glass as $\frac{3}{2}$, flint glass as $\frac{8}{5}$, and diamond as $\frac{5}{2}$.

512. To Draw the Refracted Ray. — Let MN (Fig. 252) separate the two media, air and water; and let BA be a ray of light incident at A; it is required to draw the refracted ray. With A as a centre and radii whose ratio is $\frac{4}{3}$, the index of refraction, draw two concentric circles. Through E, one of the intersections of BA with the *smaller* circle, draw EC parallel to the normal AD, cutting the larger circle at C. Draw AC. This will be the refracted ray, because $\dfrac{BF}{CG} = \dfrac{4}{3}$, a fact easily shown by geometry.[1]

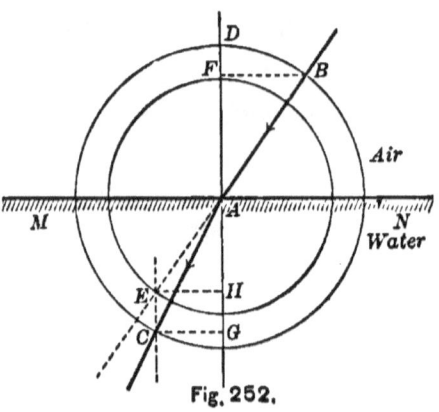

Fig. 252.

When the ray passes into a less refractive medium, through one of the intersections of the incident ray with the *larger* circle draw a parallel to the normal, cutting the smaller circle. The line through this point and the point of incidence is the refracted ray.

EXERCISES.

1. Trace a ray from air into glass, and conversely.
2. Trace a ray from water into air.
3. Trace a ray from air into diamond.
4. Trace a ray from water into glass.

513. Refraction through a Plate with Parallel Faces.
— **Exp.** — Draw a straight line on a sheet of paper. Place a piece of

---

[1] The triangles EAH and ABF are similar. Hence, $\dfrac{BF}{EH} = \dfrac{AB}{AE}$. But $EH = CG$, and $\dfrac{AB}{AE} = \dfrac{4}{3}$. Therefore, $\dfrac{BF}{CG} = \dfrac{4}{3}$.

thick plate glass on the paper so as to cover a portion of the line. On looking obliquely through the glass the line will appear broken at the edge of the plate, the part under the glass appearing laterally displaced.

To explain this, let MN (Fig. 253) represent a plate of glass, and AB a ray of light incident obliquely upon it. If the path of the ray be determined, the emergent ray will be parallel to the incident ray. Hence, the apparent position of an object viewed through a plate is at one side of its true position, a fact usually expressed by saying that the ray has suffered *Lateral Aberration.*

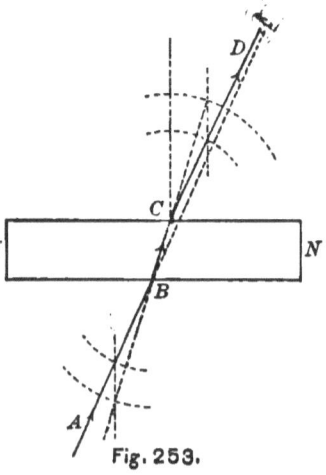

Fig. 253.

514. A Prism is that portion of a transparent substance lying between two intersecting planes. The angle between these planes is the *Refracting Angle* of the prism.

Let ABC (Fig. 254) represent a section of a glass prism made by a plane perpendicular to the refracting edge A.

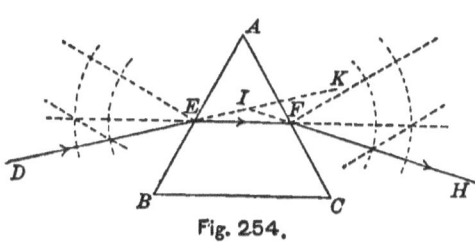

Fig. 254.

Also let DE be a ray incident on the face BA. This ray will be refracted along EF (512), and entering the air at the point F will be refracted again, taking the direction FH.

An inspection of the figure shows that the light is refracted toward the thicker portion of the prism, and consequently the apparent position of an object seen through it is displaced toward the refracting edge.

515. The Angle of Deviation. — **Exp.** — Make a small hole in the centre of a sheet of black cardboard (Fig. 255). Place in front of the cardboard a lighted spirit-lamp, with common salt sprinkled on the wick to produce a yellow flame. Arrange the flame at the same height above the table as the aperture in the screen. Behind the screen support a second one and mark on it the position of the spot of light. Now place a prism behind the hole, with the refracting edge horizontal and parallel to the screen. The position of the spot of light is different. The angle formed by the lines joining the two positions of the spot with the hole is roughly the *Deviation*.

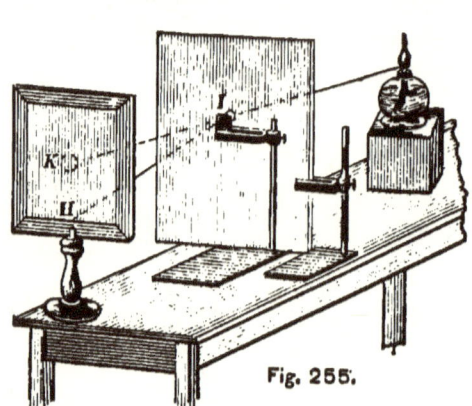

Fig. 255.

The Angle of Deviation is the angle included between the incident and the emergent ray, as *KIH* (Fig. 254). By tracing different rays through the prism, and varying the angle of the prism, the deviation will be found to vary with the angle of the prism, with the index of refraction, and with the angle of incidence. *The Least Deviation* for any given prism takes place when the angles of incidence and of emergence are equal.

516. Cause of Refraction. — Experiments show that light travels with a smaller velocity in highly refractive substances like glass than it does in air. Let a ray, *AB*, (Fig. 256) issue from a luminous point, so far distant that the waves may be considered as straight-fronted, and let it be incident obliquely on a piece of glass. A series of parallel lines perpendicular to AB will represent the waves. One part of a wave, as f, will reach the glass before other parts, and on entering the glass will travel less rapidly. The other portions on entering the glass will be retarded in succession, the result being that the wave is swung around; that is, the direction of propagation BC, perpendicular to the wave-front, is changed, or, in other words, the ray is refracted.

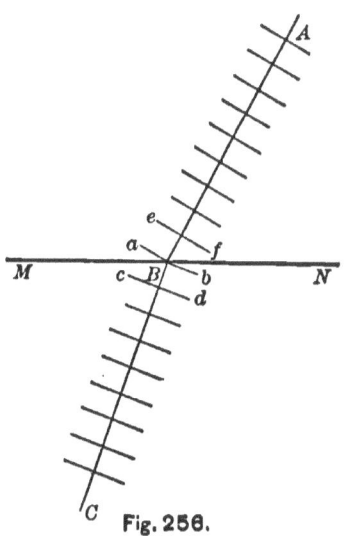
Fig. 256.

517. Phenomena of Refraction. — Since light is refracted from the normal in passing from water into air, it follows that if we look obliquely at any object submerged in water it will appear elevated above its true position. A familiar instance is to put a coin in an empty basin and place the eye in such a position that the coin cannot be seen over the side of the vessel (Fig. 257). On filling the basin with water the coin becomes visible. The apparent shoaling in the distance of a pond of water is explained in the same way. The apparent depth of water can not exceed ¾ the real depth.

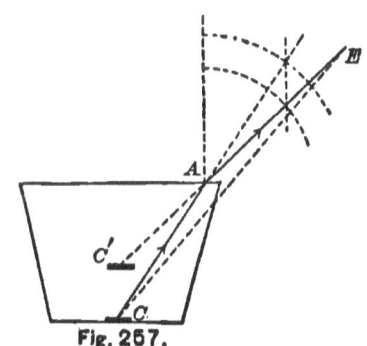
Fig. 257.

Since the lower layers of air are denser than the upper, a ray of light coming to the eye from a star will be gradually bent, describing a curve (Fig. 258). Then the direction in which the star is seen will be that of a tangent to the curve at the eye. The effect is to make the star appear higher up than its true position, making it possible to see it even when it is below the horizon.

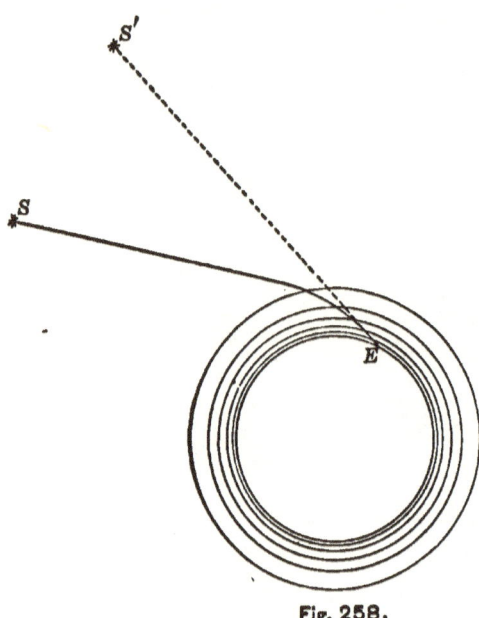

Fig. 258.

518. Total Reflection.—Exp.—Using the apparatus of Art. 510, place the cardboard against the end so that the slit is close to the bottom of the jar (Fig. 259). Reflect a beam of light upon a mirror lying on the table close to the slit, at such an angle that it passes through the slit into the liquid and is incident on the surface of the water back of the centre of the circle. Notice that the angle of refraction is greater than that of incidence. Now gradually increase the angle of incidence. The angle of refraction will increase somewhat faster than the angle of incidence. At one position for the incident ray the refracted ray coincides with the surface, making the angle of re-

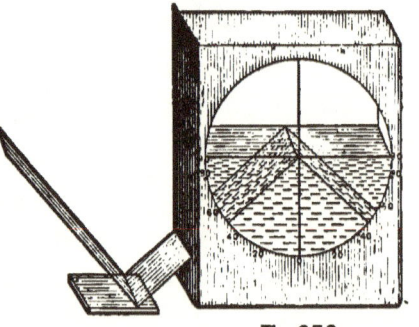

Fig. 259.

fraction a right angle. On passing this position the ray does not issue from the water, but is all reflected as if the surface were a perfect mirror.

When the angle of refraction is a right angle, the angle of incidence is called the *Critical Angle*. This angle varies with the substance, being about 48½° for water, 41° for crown glass, and 24° for diamond. When the angle of incidence exceeds the critical angle, the light suffers *Total Internal Reflection*.

519. Construction for the Critical Angle. — Let MN (Fig. 260) separate two media, as air and glass. With A as a centre draw two concentric circles, the ratio of their radii being the index of refraction, $\frac{3}{2}$. The issuing ray must lie in AN. Hence AN must pass through that point of the smaller circle determined by the normal to MN drawn through the point in which the incident ray produced cuts the larger circle (512). C is that point, and the normal CE through it cuts the larger circle at E. Draw EA and produce it to B. BA is the incident ray, and BAD is the critical angle.

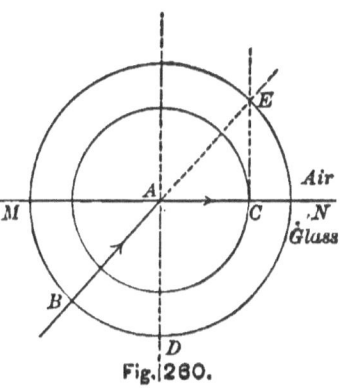

Fig. 260.

520. Illustrations of Total Reflection. — Of all the rays diverging from a point at the bottom of a pond and incident on the surface, only those within the cone whose semi-angle is 48½° can pass into the air, on account of the separation of the refracted from the totally reflected rays by total internal reflection (Fig. 261). Hence, an observer under water sees all objects without as if they were crowded within this cone; and beyond this cone, he sees by reflection objects lying on the bottom of the pond.

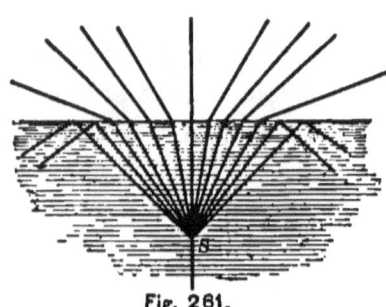
Fig. 261.

If a test-tube, part full of mercury, be held obliquely in water, and looked at from above, the empty part will appear much brighter than the filled part, since the former is seen by light totally reflected, whereas the latter is seen by ordinary reflection.

Total reflection in glass is best shown by means of a prism whose cross-section is a right-angled isosceles triangle (Fig. 262). A ray incident normally on either face about the right angle enters the prism without refraction and is incident on the hypothenuse at an angle of 45°, which is greater than the critical angle (518). Hence the ray suffers total reflection and leaves the prism at right angles to the incident ray. Such a prism makes the best possible reflector in any optical instrument where it is desirable to change the direction of the light by 90°.

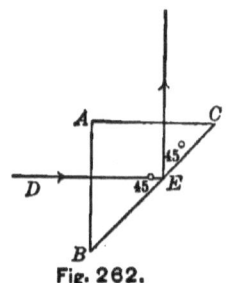
Fig. 262.

EXERCISES.

1. Construct the critical angle for water.
2. Construct the critical angle for diamond.
3. If you hold a glass of water with a spoon in it, a little above the level of the eye, and look upwards at the under surface of the water, the lower part of the spoon is seen reflected in it. Explain.

LIGHT. 339

4. When glass is pounded into small particles it appears white and loses its transparency. Why?

5. Draw a prism with a refracting angle of 60°. Let the angle of incidence for a ray be 15°. Trace the ray through the prism, assuming the index to be 1.5.

6. Why does a sphere appear like a spheroid when under water?

VI. LENSES.

521. A **Lens** is a portion of a transparent substance bounded by two curved surfaces, or one curved and one

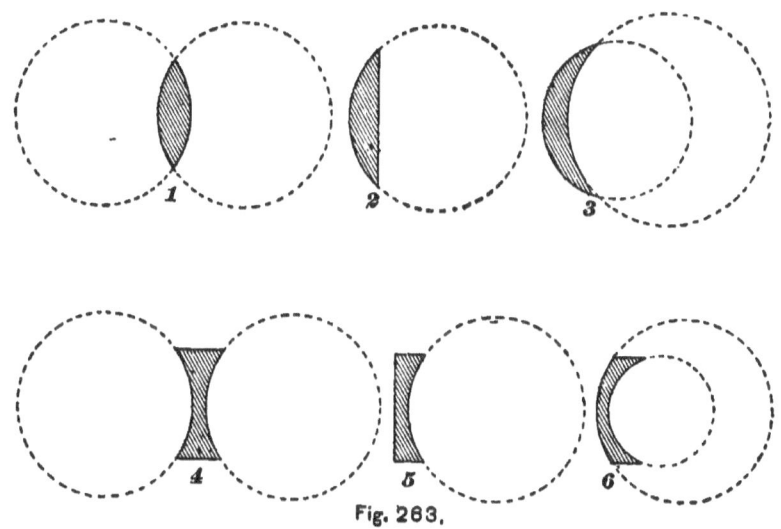

Fig. 263.

plane surface. Those most commonly met with are portions of spheres and are classified as follows:

1. Double-Convex, — both surfaces convex.
2. Plano-Convex, — one surface convex, one plane.
3. Concavo-Convex, — One surface convex, one concave.

} Converging lenses, thicker at the middle than at the edges.

4. Double-Concave, — both surfaces concave.
5. Plano-Concave, — one surface concave, one plane.
6. Convexo-Concave, — one surface concave, one convex.

} Diverging lenses, thinner at the middle than at the edges.

The double-convex lens may be regarded as the type of the converging class, and the double-concave lens of the diverging class, since the properties of each of these lenses apply to all those of the same class.

522. Terms Defined. — The centres of the spherical surfaces bounding a lens are the *Centres of Curvature*.

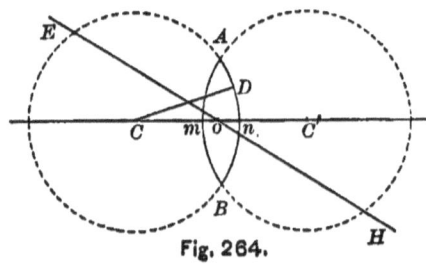

Fig. 264.

The *Principal Axis* is the straight line passing through the centres of curvature CC' (Fig. 264). In the plano-lenses there is but one centre of curvature, the principal axis being the normal to the plane surface through that centre. The normal at any point of the surface of a lens is the radius of the sphere drawn to that point. Thus CD is the normal to the surface AnB at D.

523. Optical Centre. — It is shown in geometrical optics that for every lens there is a point having the property that any ray passing through it and the lens suffers no change of direction. This point is called the *Optical Centre*. In lenses whose surfaces are of equal curvature, it is their centre of volume; in plano-lenses, it is the intersection of the principal axis with the curved surface. Any straight line, as *EH* (Fig. 264), through the optical centre is a *Secondary Axis*.

524. To Trace a Ray through a Lens.—Let *MN* represent a lens whose centres of curvature are *C* and *C'*, and

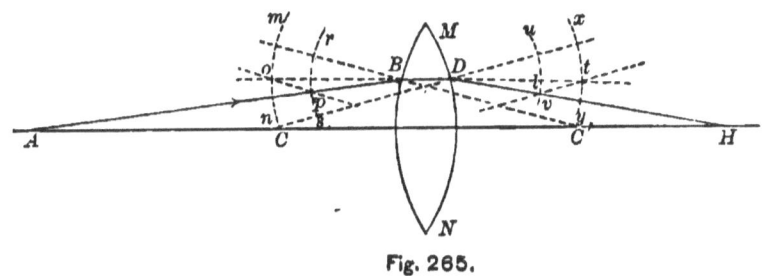

Fig. 265.

AB the ray to be traced through it (Figs. 265, 266). Draw the normal, *C'B*, to the point of incidence. With *B* as a centre, draw the arcs *mn* and *rs*, making the ratio of their radii equal the index of refraction, $\frac{3}{2}$.[1] Through

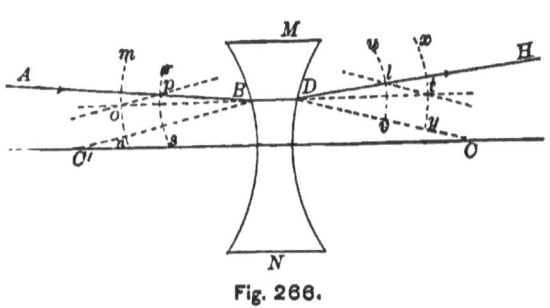

Fig. 266.

p, the intersection of *AB* with *rs*, draw *op* parallel to the normal, *C'B*, and cutting *mn* at *o*. Through *o* and *B* draw *oBD*; this will be the path of the ray through the lens (512). At *D* it will again be refracted; to determine the amount, draw the normal *CD* and the auxiliary circles, *xy* and *uv*, as before. Through

---

[1] In all the figures which follow, the index is taken as $\frac{3}{2}$.

t, the intersection of *BD* produced with *xy*, draw *lt* parallel to the normal *CD*, cutting *uv* at *l*. Through *D* and *l* draw *DH*; this will be the path of the ray after emergence. [Compare this mode of procedure with that of Art. 512.] It should be noticed that the convex lens bends the ray toward the principal axis, whereas the concave lens bends it away from it.

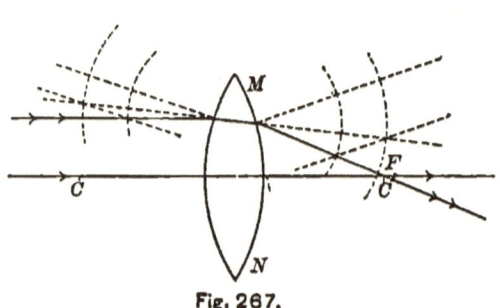

Fig. 267.

525. Foci of Convex Lenses. — A clear understanding of this subject may be reached through the consideration of the following cases:

First. When the incident rays are parallel.

An inspection of Fig. 267 shows that the focus is real.

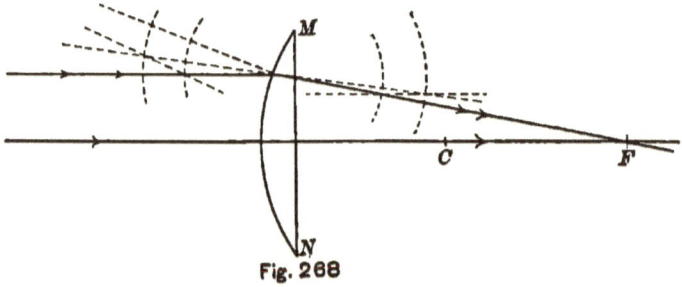

Fig. 268.

Since the rays are parallel, the focus is known as the *Principal Focus*, and its distance from the lens is the *Focal Length*. When the index is taken as $\frac{3}{2}$, it will be observed that in the case of the double convex lens, the principal focus is at the centre of curvature, the focal length being the radius of curvature, and that in the plano-convex the focal length is twice the radius of curvature. If the index were greater, the focal length would be shorter, and conversely.

Second. When the incident rays diverge from a point more than twice the focal distance from the lens.

In Fig. 269, AD and AB are two rays diverging from A.[1] They focus at less than twice the focal distance.

Third. When the incident rays diverge from a point at twice the focal distance from the lens.

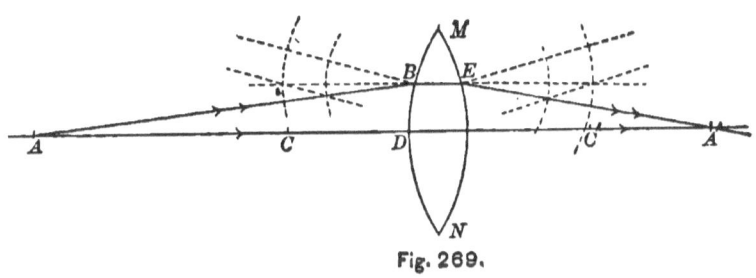

Fig. 269.

[Let the student draw a figure and show that the focus is real and at twice the focal distance.]

Fourth. When the incident rays diverge from a point at less than twice the focal distance.

[Let the student draw a figure and show that the focus is real and at more than twice the focal distance.]

Fifth. When the rays diverge from the principal focus.

[Let the student draw a figure and show that the emergent rays are parallel.]

Sixth. When the rays diverge from a point between the focus and the lens.

---

[1] Since intermediate rays focus at the same point, it is necessary to trace but two rays in any of these figures. For simplicity, the principal axis is taken as one of them.

Fig. 270 shows that the divergence of the rays is not wholly overcome by the lens, but that they leave the lens as if they emanated from a point farther from the

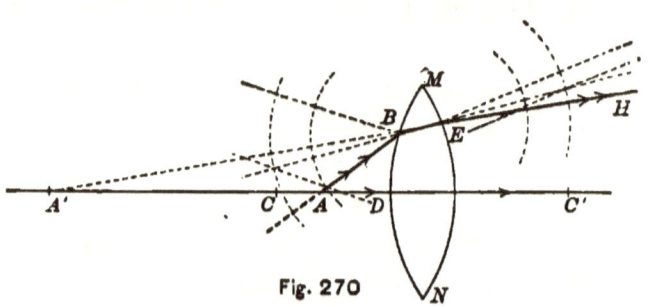

Fig. 270

lens than the actual radiant point. Hence, the focus is virtual.

Seventh. When the rays converge to a point beyond the lens.

[Let the student draw a figure showing that they focus at a point nearer the lens than the principal focus.]

526. Foci of Concave Lenses. — The following cases may be considered:

First. When parallel rays are incident on the lens.

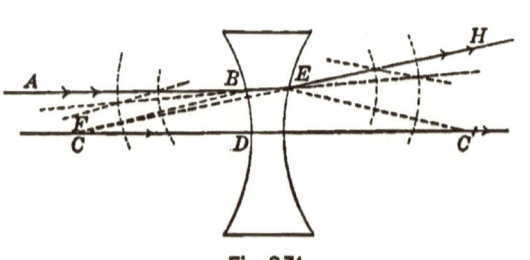

Fig. 271.

Fig. 271 shows that the rays diverge on leaving the lens as if they came from a point. Hence, the principal focus is virtual. When the index of refraction is $\tfrac{3}{2}$ the focal length equals the radius of curvature in the case of the double-concave lens, and twice the radius in the case of the plano-concave lens.

Second. When diverging rays are incident on the lens. Fig. 272 shows that the focus is virtual, situated between the principal focus and the lens, and moves nearer to the lens as the divergence increases.

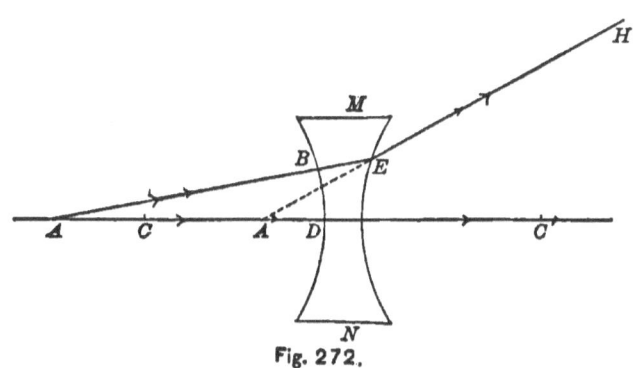

Fig. 272.

Third. When converging rays are incident on the lens. A comparison of Figs. 273, 274, and 275 makes it evident that the degree of convergence affects the general character of the results. Fig. 273 shows that a pencil, as $ABCD$, converging to the principal focus F, emerges with its rays parallel. Fig. 274 shows that a pencil, as $ABCD$, converging to a point on the axis nearer to the lens than the principal focus, as K, emerges less converging, focusing beyond F at A'. Fig. 275 shows that a pencil, as

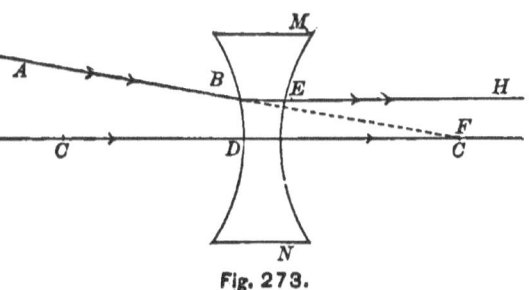

Fig. 273.

$ABCD$, converging to a point beyond the principal focus, as K, emerges as a diverging pencil having a virtual focus at A'.

527. The General Effect of Lenses on light may be summarized as follows: $\left.\begin{array}{c}\text{Concave}\\\text{Convex}\end{array}\right\}$ lenses $\left\{\begin{array}{c}\text{increase}\\\text{lessen}\end{array}\right\}$ the divergence of diverging pencils and $\left\{\begin{array}{c}\text{lessen}\\\text{increase}\end{array}\right\}$ the convergence of converging pencils.

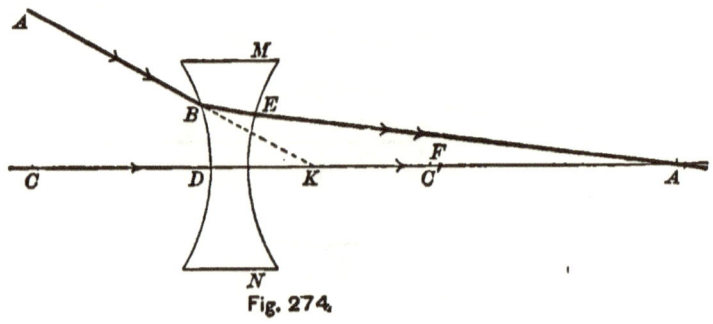

Fig. 274.

528. Formation of Images. — The image of an object formed by a lens may be found by means of the images of its points. The work is considerably lessened by observing the following rule:

Draw secondary axes through the extremities of the object. These will be rays of light which suffer no change

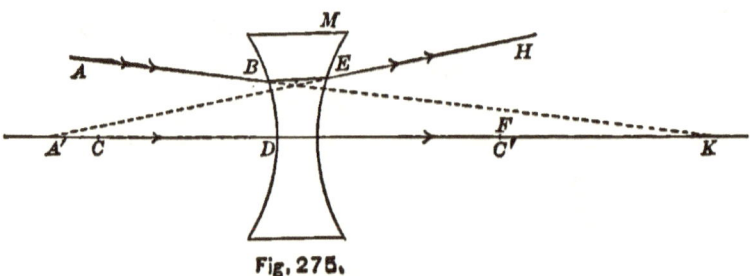

Fig. 275.

of direction. (Why?) Also through these extremities draw rays parallel to the principal axis, and find by construction the amount they are refracted (524) on entering the lens. On leaving the lens they will pass through the principal focus. (Why?) The images of these extremities will be the intersections of the rays drawn from them. The image of any point is always on the secondary axis passing through it.

To illustrate, let *AB* be the object and *MN* the lens (Fig. 276). Rays along the secondary axes pass through the lens without deviation. The rays *AD* and *BH*, parallel to the principal axes, are refracted on entering the lens along *DE* and *HI* respectively, and pass through *F*, the principal focus, after leaving the lens. The intersec

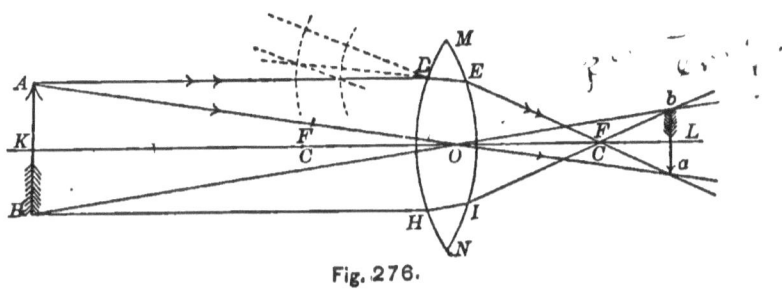

Fig. 276.

tion of *Aa* with *ADEa* is the image of *A*, and that of *BHIb* with *Bb* is the image of *B*. If other rays from *A* and *B* be drawn, they will focus at *a* and *b* respectively. Hence, *ab* is the image of *AB*.

The student should draw figures and establish the following propositions for converging lenses:

I. *When the object is at an infinite distance, that is, when the rays are parallel, they converge to the principal focus, making the image a point.*

II. *When the object is at a finite distance greater than twice the focal distance, the image is real, inverted, situated beyond the principal focus, and is smaller than the object.*

III. *When the object is at twice the focal distance, the*

image is real, inverted, situated at twice the focal distance, and is of the same size as the object.

IV. *When the object is at less than twice the focal distance, the image is real, inverted, situated beyond the focus, and is larger than the object.*

V. *When the object is at the focus, the light leaves the lens in parallel lines, no image being formed.*

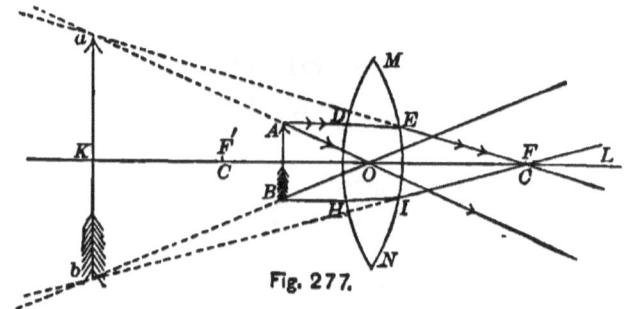
Fig. 277.

VI. *When the object is between the focus and the lens, the image is virtual, erect, and enlarged.*

The images formed by concave lenses are always virtual, erect, and smaller than the object. The distance of the object from the lens affects only the size of the image.

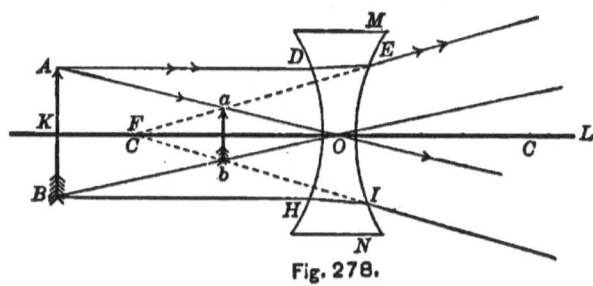
Fig. 278.

529. Experimental Illustrations. — **Exp.** — Let the rays of the sun fall on a convex lens. Hold beyond the lens a sheet of paper, moving it till the round spot of light is smallest and brightest. If held steadily a hole may be burned in the paper. This spot is an image of the sun; and because the solar rays are parallel, it marks the principal focus of the lens.

The experiment illustrates a simple method of finding the focal length of a convex lens, and also proves that radiant heat like light can be refracted. Convex lenses are often called "burning-glasses," because of their power to focus the "heat rays."

Exp. — Arrange in a straight line on the table a lamp, a convex lens, and a screen. Give the lamp successively the positions described in the second, third, and fourth cases, adjusting the screen each time till a sharply defined image is obtained. The results will be in harmony with the conclusions already given. If the lamp is placed at the focus of the lens, only a blurred image is obtained; if between the focus and the lens, no image is formed on the screen, but a magnified image of the lamp is seen on looking through the lens.

If after focusing the image on the screen, the eye be placed in line with the object and lens, and the screen be removed, an inverted image of the object can be seen suspended in mid-air. The projection of images on a screen by the aid of a lens has its application in the optical lantern, the compound microscope, the telescope, and the photographer's camera. The formation of virtual images by a convex lens is applied in the simple magnifier and in the eye-pieces of telescopes and compound microscopes.

530. Measurement of Focal Length. — **Exp.** — Support the perforated screen of Art. 505 in a vertical plane, and parallel to it place the convex lens and a second screen. The centre of the lens should be in a horizontal line with the centre of the aperture. Place a lamp behind the first screen to illuminate the cross-threads. By trial find a position for the lens and the second screen such that the image of the aperture is of the same size as the aperture itself and very clearly defined. Then the distance between the two sheets of cardboard is four times the focal length of the lens. Why?

850   ELEMENTS OF PHYSICS.

531. Spherical Aberration. — If rays from any point be drawn to different parts of a lens, and their directions be determined after refraction, it will be found that those incident near the edge of the lens cross the principal axis, after emerging, nearer the lens than those incident near the middle. [Let the student draw a figure proving this.] This indefiniteness of focus is called *Spherical Aberration*, the effect of which is to lessen the distinctness of images formed by the lens. In practice, annular screens, called *Diaphragms*, are used to cut off the marginal rays, thereby rendering the image sharper in outline, but less bright. In the large lenses used in telescopes, the curvature of the lens is made less toward the edge, so that all the rays are brought to the same focus.

VII. DISPERSION.

532. Analysis of White Light. The Solar Spectrum.
— **Exp.** — Admit a beam of sunlight into a dark room through a

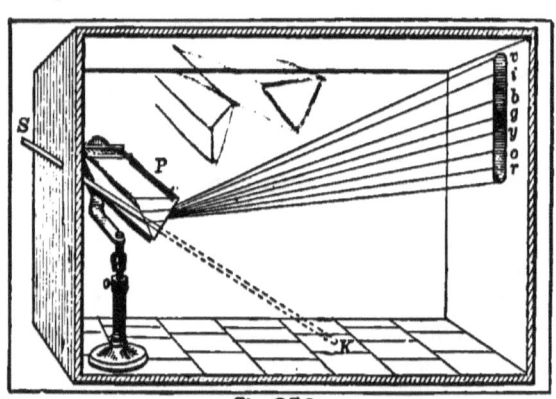

Fig. 279.

small hole in the shutter (Fig. 279). A colorless image of the sun is formed on the floor or wall. In the path of the beam place a triangular glass prism so as to refract the light. Instead of a colorless image being formed, there is now a rainbow-colored band gradually changing from red at one end, through orange, yellow, green, blue, indigo, to violet at the other end.

This experiment was first explained by Sir Isaac Newton in 1666. It shows that white or colorless light is a mixture of an infinite number of differently colored rays, differing in refrangibility, the red being least and the violet most refrangible. This many-colored image is called the *Solar Spectrum*, and the opening out or separating of the beam of white light is called *Dispersion*.

533. Synthesis of Light. — Exp. — Project a spectrum of sunlight on the screen. Now place a second prism like the first behind it but reversed in position (Fig. 280). There will be formed a colorless image, slightly displaced. The second prism reunites the colored rays, making the effect that of a thick plate of glass. The recomposition of the colored rays into white light may also be effected by receiving them on a concave mirror or a large convex lens.[1]

Fig. 280.

534. Chromatic Aberration. — Exp. — Reflect a beam of sunlight into a dark room through a small hole in the shutter. Project an image of the aperture on the screen, using a double-convex lens for the purpose. The round image will be seen to be bordered with the spectral colors.

This defect is known as *Chromatic Aberration*. It is caused by the lens refracting the colored rays to different foci. The violet rays being more refrangible than the red will have their focus nearer to the lens in consequence. By examining Fig. 281 it will be seen that if

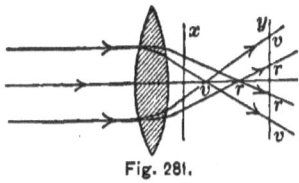

Fig. 281.

---

[1] A glass cylinder filled with water makes an excellent substitute for the lens.

a screen were placed at x the image would be bordered with red, and if at y with violet.

535. The Achromatic Lens.—Exp.—Project a spectrum of sunlight on the screen, using a prism of crown glass, and note the length of the spectrum. Repeat the experiment with a prism of flint glass having the same angle. The spectrum formed by the flint glass is seen to be about twice as long as that given by crown glass, whereas the position of the middle of the spectrum on the screen is about the same in the two cases. Now use a flint glass prism whose refracting angle is half that of the crown glass one. The spectrum resembles in size that given by the crown glass prism, but the deviation of the middle of it is considerably less. Finally, place this flint glass prism in a reversed position against the crown glass one. The image of the aperture is no longer colored, but the deviation is about half that of the crown glass alone.

The above facts suggested to Dolland, an English optician, in 1757, that by combining a double-convex lens of crown glass with a plano-concave lens of flint glass, the dispersion by the one would neutralize that by the other, whereas the refraction would be reduced about half (Fig. 282). Such a combination of lenses is said to be *Achromatic*, since images formed by them are not fringed with the spectral colors.

Fig. 282.

536. The Rainbow.—Exp.—Fill an air-thermometer bulb, about 4 cm. diameter, with clear water. Reflect a beam of sunlight into a dark room, passing it through a circular opening, about 3.75 cm. diameter, cut in a large sheet of white cardboard. Support the bulb so that the cylindrical beam is incident on it. There will appear on the cardboard screen one or more circular spectra, resembling rainbows.

The Rainbow is a solar spectrum formed by spherical rain-drops, after the manner of the above experiment.

Usually two bows are visible, the *Primary* and the *Secondary*. The *Primary bow* is the inner and brighter one, and is distinguished by being red on the outside and violet on the inside. *The Secondary bow* is much fainter, and has the order of colors reversed.

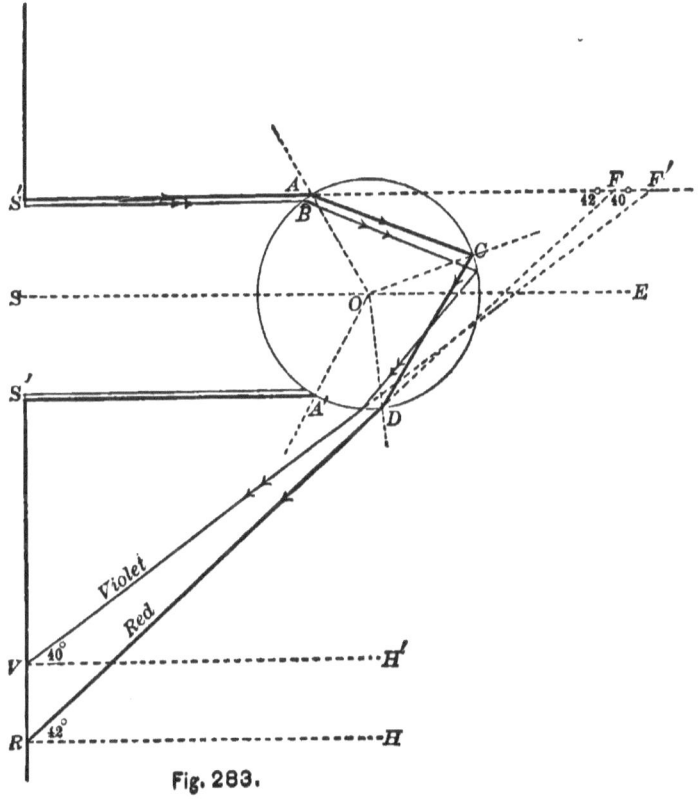

Fig. 283.

537. The Experiment Explained. — When light is incident on the surface of any transparent medium, part of it passes into the medium, and part is reflected. Hence, in the case of the sphere of water, part of the beam incident on it is reflected, and part passes within. Of that which enters the globe, part passes through it and part is reflected from the back surface. It is this reflected part which forms the image on the cardboard screen.

Let the circle whose centre is O (Fig. 283) represent the globe of water, and S, S', rays of sunlight incident upon it. Of all the rays entering the globe, it is shown in mathematical optics that the *red* rays incident in the immediate vicinity of 59° 23′ 30″ from the axis SE, as at A,

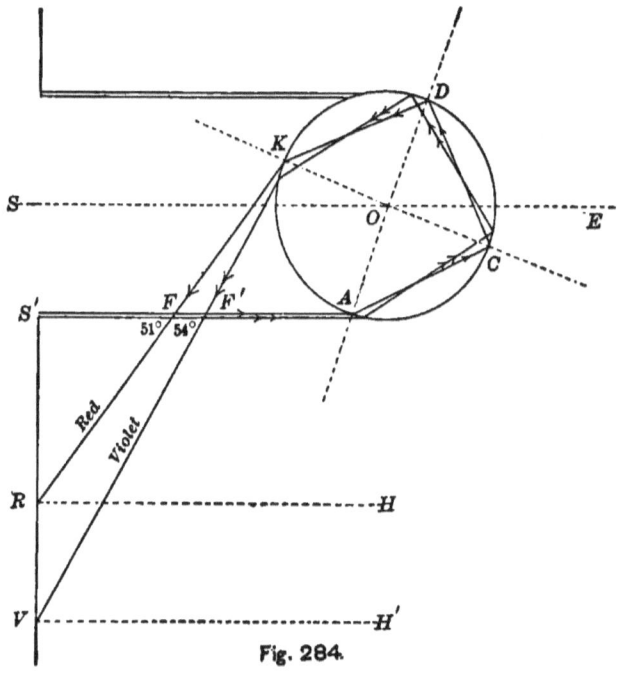

Fig. 284.

keep together after reflection and subsequent refraction, that is, are parallel on leaving the sphere, and hence have sufficient intensity to produce a colored image on the screen; and the *violet* ones at 58° 40′ from the axis, as at B, produce a violet image for a like reason. Between these positions, the other colors arrange themselves. Since the sphere is symmetrical with respect to SO, it follows that the colored images will appear as circles. In this way is produced the inner colored ring in the above experiment. The angle, $S'FR$, is the deviation of the red rays, and equals 42° 1' 40''; the deviation of the violet rays, $S'F'V$, is 40° 17'. Hence, the red circle on the screen has a radius of about 42°, and the violet one of about 40°.

The second colored band is caused by rays twice internally reflected, as shown in Fig. 284. The violet rays incident at 71° 49' 55'' from the axis will emerge parallel, and produce the violet part of the image; and the red ones incident at 71° 26' 10'' will form the red part for a like reason. The deviation, $S'F'V$, of the violet rays is 54° 9' 20'', and of the red, $S'FR$, is 50° 58' 50''. Hence, the second colored band has a radius of about 54° for the violet part, and 51° for the red.

538. The Primary Bow Explained. — Every rain-drop acts on sunlight in exactly the same manner as the sphere of water in the above experiment. An observer at E (Fig. 285) with his back to the sun receives red light from every rain-drop situated 42° from the line SC, drawn through the sun and the eye, and violet from those 40° distant. The drops, being approximately spheres, send out light on all sides, and hence all of

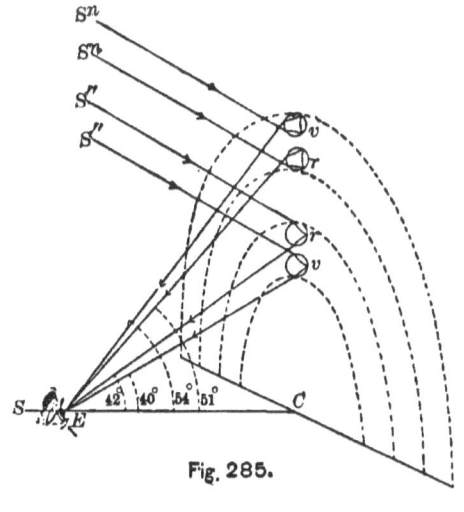

Fig. 285.

those situated in a circle about C and distant 42° will send red light to the eye, and give the appearance of a red

356   *ELEMENTS OF PHYSICS.*

circle, while those at 40° will send violet light, the other colors arranging themselves between these limits.

539. The Secondary Bow is due to two reflections within the rain-drops; and in order for the light to reach the eye, these drops must be distant from C about 54° for the violet and 51° for the red. The diminished brightness is caused by the loss of light attending each reflection.

VIII. COLOR.

540. The Color of light depends on its wave-length, red being due to the longest waves and violet to the shortest. The following table expresses the wave-lengths corresponding to the centre of each color in the solar spectrum in ten-millionths of a millimetre:

Red,	7000	Yellow,	5808	Indigo,	4383
Orange,	5972	Green,	5271	Violet,	4059
		Blue,	4960		

In white light the number of colors is infinite, and they pass into one another by imperceptible gradations of shade and refrangibility. Color stands related to light in the same way that pitch does to sound. In artificial light, certain colors are either feeble or wanting, as can be proved by an examination of their spectra. Hence, artificial lights are not white, but each one is characterized by the color that predominates in its spectrum.

541. The Color of Opaque Bodies. — **Exp.** — Paste a strip of white paper, say, 3 cm. long and 2 mm. wide, on a piece of black cardboard several times larger. View this strip, placed in a strong light, through a glass prism, holding its edges parallel to the length of the strip. The image is a spectrum, colored like that produced by sunlight. If a red strip of paper similarly mounted is examined in the same way, the spectral image is red at one end, while the colors belonging to the other end are dim or absent. In like manner if a blue strip is examined, the spectral image is characterized by blue, the other colors being mostly wanting.

The experiment shows that the color of opaque bodies is due to the kind of light they reflect; that a body is red because it reflects chiefly, if not wholly, the red rays of the light incident upon it, the others being absorbed wholly or partly at its surface; and white, if it reflects all the rays in about equal proportions.

542. **Color not of the Body.** — **Exp.** — Project the solar spectrum on a white screen. (Why white?) Hold pieces of colored paper or ribbon successively in different parts of the spectrum. The red pieces appear brilliantly red in the red part of the spectrum and black in most of the others, a blue piece is blue only in the blue part of the spectrum, and a black piece is black, no matter where it is placed.

The experiment shows that bodies have no color of their own, but change with the nature of the incident light. This truth is illustrated by the difficulty experienced in matching colors by artificial lights, and by the changes in shade some colored fabrics undergo when taken from sunlight into gaslight. Artificial lights are largely deficient in blue and violet rays; and hence all complex colors into which blue or violet enters, as purple and pink, change their shade when viewed by them.

543. **Color of Transparent Bodies.** — A transparent body is colorless when it absorbs no light, or absorbs all colors in like proportion. If it absorbs more of one color than another, its color is that which results from combining the transmitted rays.

358   ELEMENTS OF PHYSICS.

544. Mixing Colors. — **Exp.** — Cut out of colored paper some disks, about 15 cm. in diameter, with an opening at the centre for mounting them on the spindle of a whirling machine (Fig. 287) or for slipping them over the handle of a heavy spinning top. Slit them along a radius from the circumference to the centre, so that two or more of them can be placed together, exposing any proportional part of each one as desired. Select seven disks, whose colors most nearly represent those of the solar spectrum; put them together so that equal portions of the colors are exposed. Clamp on the spindle of the whirling machine and rotate them rapidly, or slip them over the handle of a heavy spinning top. When viewed in a strong light the color is nearly white.

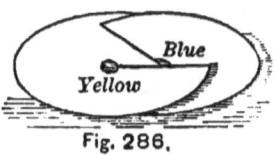
Fig. 286.

Fig. 287.

This method of mixing colors is based on the physiological fact that a sensation lasts longer than the impression producing it. Before the sensation due to one impression has ceased, the disk has moved so that a different impression is produced. The effect is equivalent to superposing each color on all the others, as was done when the recomposition of white light was effected by a lens or mirror (533).

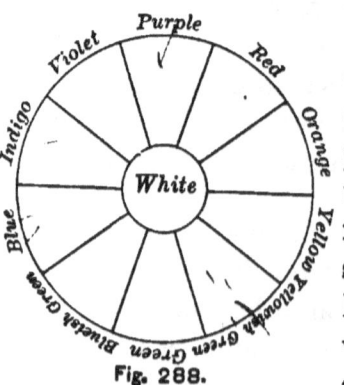
Fig. 288.

If red, green, and violet disks, or red, green, and blue ones, are used, exposing equal portions, nearly white is obtained on rotating them rapidly. If any two colors standing opposite each other in Fig. 288 are used the result is white; and if any two alternate ones are used the result is the intermediate one. By using the red, the green, and the violet disks, and exposing different proportions, it has been found possible to produce any color of the spectrum. This fact suggested to Dr. Young the theory that the eye is sensitive to only three colors, and that our recognition of different colors is due to the excitation of these three in varying degrees (562).

545. **Complementary Colors** are any two colors which produce white light on mixing. Thus, the colors red and bluish-green are complementary; also orange and light-blue. The placing of such colors adjacent to each other causes a heightening of each, as in the case of red and green; two non-complementary colors placed adjacent cause a toning down of each.

546. **Mixing Pigments.** — **Exp.** — Draw a broad line on a slate or black-board with a yellow crayon. Over this draw a line with a blue crayon. The result will be a band distinctly green. The explanation is that the yellow crayon reflects green light as well as yellow, and absorbs all the other colors. The blue crayon also reflects the green light, but absorbs others except the blue. In mixing the crayons, the product absorbs all but the green, and hence the mark on the board is green. Another illustration of this principle is seen in the mixing of a solution of potassium chromate and an ammoniacal solution of copper sulphate. The former is yellow, the latter is blue, but on mixing the solution is green. (Why?)

It must be borne in mind that these statements apply to colored substances and not to colored lights; for if we mix the spectral colors blue and yellow, the product is white instead of green.

IX. SPECTRUM ANALYSIS.

547. A Pure Spectrum. — Exp. — Tack across the opening of the porte lumière [1] a piece of tin in which is cut a very narrow slit about 3 cm. long. Project, with a convex lens of about 30 cm. focal length, an image of this slit on a distant white screen. Close to the lens, between it and the screen, place a prism of flint glass, 60° angle, or, what is better, one of carbon bisulphide, and change the position of the screen, keeping it at the same distance from the lens, to a position where the refracted pencil will fall normally upon it. Turn the prism around its vertical axis till the least deviation is secured. The colors will now show with much greater distinctness than in the previous projections of the spectrum where a round aperture was used.

By using a narrow slit, we obtain a spectrum in which there is little or no overlapping of the colored images, be-

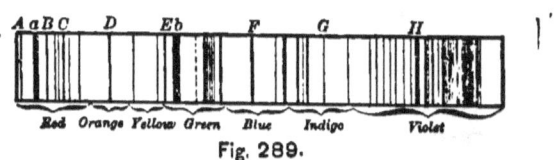

Fig. 289.

cause they are rectangular in shape and very narrow. The result is, therefore, a *Pure Spectrum*, one made up of a *succession* of colored images of the rectangular slit.

548. The Fraunhofer Lines. — When a pure spectrum is obtained by careful focusing it will be found crossed by a number of parallel dark lines (Fig. 289). This discovery was made by Wollaston in 1802. In 1814, Fraunhofer, of Munich, by viewing the spectrum through

---

[1] The porte lumière is a device by means of which a beam of sunlight can be reflected horizontally into a darkened room through an opening in the shutter. In its simplest form it is a hinged mirror set outside of the window, and so arranged that its position can be changed from within.

a telescope, counted about 600 lines, and mapped the position of nearly two-thirds of them, designating the more prominent ones by the letters A, B, C, D, E, F, G, H. Since Fraunhofer was the first to study carefully the position of these lines, it is customary to refer to them as the Fraunhofer lines. In recent times, their number has been greatly extended through the investigations of Lockyer, Rowland, and Langley.

Since these lines do not change their position with reference to the color in which they occur, they are of great value as reference marks. For example, in measuring the index of refraction of a substance, owing to dispersion, it is necessary to select some particular color or line of the spectrum for the refracted ray. Since the E-line in the green marks approximately the middle of the visible spectrum, it is customary to make all determinations with reference to it. Sometimes the D-line in the yellow is chosen, on account of its greater conspicuousness. The index of refraction of a substance, to be exact, should always be given for some particular spectral line.

549. The Spectroscope (Fig. 290) is an apparatus for viewing spectra. In one of its simplest forms, it consists of a prism, a telescope, and a tube called the *Collimator*, carrying an adjustable slit at the outer end, and a convex lens at the other, to render parallel the diverging rays

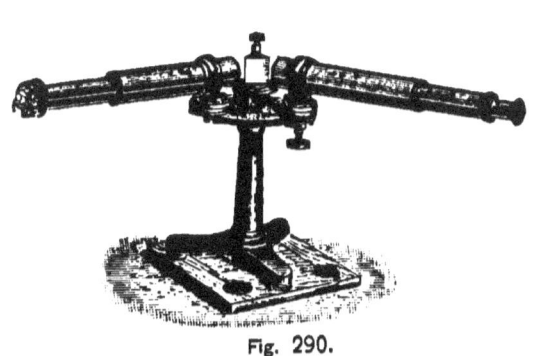
Fig. 290.

coming from the slit. To mark the position of the spectral lines, there is provided on the supporting table a divided circle which is read by the aid of the vernier attached to the telescope arm. By increasing the number of prisms, the purity of the spectrum is increased, and consequently lines previously concealed by the overlapping of the images of the slit are brought to view.

550. Continuous Spectra. — Exp. — Place a lighted candle or lamp in front of the slit of the spectroscope. The spectrum will be seen to pass from red at one end, through the various colors, to violet at the other, without any breaks, that is, the spectrum is not crossed by dark lines or bands.

Such spectra are called *Continuous Spectra*.

551. Discontinuous Spectra. — Exp. — Place a Bunsen flame in front of the slit of a spectroscope. Hold in the flame a strip of asbestos or a platinum wire, after dipping it into a solution of table salt. On looking through the telescope, a bright yellow line is seen instead of the rainbow band, and, furthermore, it coincides in position with the Fraunhofer D-line. If we try a lithium salt, we shall see a bright carmine line.

Spectra consisting simply of bright lines are called *Discontinuous* or *Bright-line Spectra*.

552. Absorption Spectra. — Exp. — Turn the slit of the spectroscope toward an electric light.[1] A continuous spectrum will be seen. Between the slit and the light place the flame of a spirit-lamp colored yellow by means of common salt in the wick. A dark line will now be seen crossing the yellow of the spectrum, the light at that point having been absorbed by the flame. When the electric light is shut off, the line is seen as a bright yellow one.

---

[1] In the absence of an electric light, sunlight may be used. In that case, the principle will be illustrated in the increased darkness of the D-line by the absorption due to the yellow flame. This yellow flame is caused by the sodium in the salt.

Spectra crossed by dark lines are called *Absorption* or *Reversed Spectra*. The spectra of the sun and most stars are of this class.

553. Laws of Spectra. — Experiments like the last three establish the following laws regarding the formation of spectra :

I. *All solids, liquids, and dense vapors and gases, when heated to incandescence, give continuous spectra.*

II. *All rarefied gases and vapors heated to incandescence give discontinuous or bright-line spectra.*

III. *Every gas or vapor absorbs rays of the same refrangibility as it emits when heated to incandescence.*

554. Fraunhofer Lines Accounted for.—It was shown in a previous experiment (552) that sodium vapor extinguishes the light emitted by a luminous body so as to produce a dark line in the spectrum identical in position with the D-line of the solar spectrum. Similar experiments show that every substance through which light passes absorbs rays of definite refrangibility, that is, every substance has its own absorption spectrum. These facts were first established by Kirchhoff, suggesting to him the following explanation of the Fraunhofer lines : The heated nucleus of the sun gives off light of all degrees of refrangibility. Its spectrum would therefore be continuous were it not surrounded by an atmosphere of metallic vapors and gases, which absorb or weaken those rays of which their own spectra consist. Hence, the parts of the spectrum which would have been illuminated by those particular rays have their brightness diminished, since the rays from the nucleus are absorbed and the illumination is due solely to the less intense light coming from the vapors.

These absorption lines are not lines of no light, but are lines of diminished brightness, appearing dark by contrast with the other parts of the spectrum.

555. Spectrum Analysis consists in detecting the presence of bodies by the spectra of their heated vapors. The great delicacy of the method is exhibited in the statement made by Prof. Swan, that he was able to detect by its spectrum the presence of $\frac{1}{2500000}$th part of a grain of sodium.

The applications of the spectroscope are many and various. By an examination of their absorption spectra normal and diseased blood are easily distinguished, the adulteration of substances is detected, the chemistry of the stars is determined, and many processes in the arts are made successful.

EXERCISES.

1. When is an object black? When gray?
2. What must be the position of the slit with reference to the lens in the collimator tube of the spectroscope?
3. Why are rainbows seen only in the morning and the evening?
4. Why does paint change the color of an object?

X. INTERFERENCE OF LIGHT.

556. Interference by Reflection. — It was seen in the study of sound that two wave-motions may interfere in such a way that they either partially or entirely neutralize each other. The fact that rays of light can also be made to neutralize each other affords the strongest proof of the correctness of the wave theory of light.

557. Newton's Rings. — Exp. — Press together at their centre two small pieces of heavy plate glass, using a small iron clamp for the purpose. By looking obliquely at the glass curved bands of color may be seen surrounding the point of greatest pressure. If the observation be made in a light of one color, as that given by a spirit-lamp with common salt in the wick, these curved bands will be yellow, separated by dark ones.

The experiment, of which the above is a modification, was first performed by Hooke, and afterwards repeated by Newton. Each used a plano-convex lens of long focus resting on a plate of plane glass. Fig. 291 shows a section of the apparatus. Between the lens and the plate there is a wedge-shaped film of air, very thin, quite similar to that between the two glass plates used in the above experiment. Now if light be incident on AB, a portion will be reflected from ACB, and another portion from DE. Since that reflected from DE has farther to travel by twice the thickness of the air-film than that from ACB, and the film gradually increases in thickness from C outward, it follows that at some places the two reflected portions will meet in like phase and at others in opposite phase, causing a strengthening of the light in the first case and an extinction of it in the second. If red light be used the appearance will be that of a series of concentric circular red bands separated by dark ones, each shading off into the other. If violet light be employed the colored bands will be closer together on account of the shorter wave-length. Other colors will give bands intermediate in width between the red and the violet. From this it follows that if the plates be illuminated with white light at every point some one color will be destroyed, leaving the color at that point due to the mixing of the remaining colors of

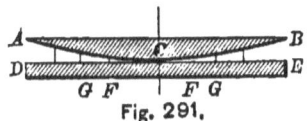

Fig. 291.

white light, and the point C will be surrounded with rainbow-like bands of color.[1]

558. Illustrations. — The colors of the soap-bubble, of oil on water, of heated metals which easily oxidize, are all due to the interference of light produced by thin films. The colors surrounding cracks in ice, and those of iridescent glass, are explained in like manner. Striated surfaces, like mother-of-pearl, some kinds of shells, and the feathers of some birds, owe their beautiful colors to the interference of light reflected from different parts of the small grooves. Gratings, consisting of parallel lines ruled on speculum metal, yield spectra by reflection in which wave-lengths are proportional to deviation.

559. Diffraction. — Exp. — Cut a slit 2 mm. wide in a black card, and hold it in front of a flame so as to be brightly illuminated. Blacken a strip of window-glass with India ink. When dry, rule with a fine needle a number of parallel lines, quite close together, cutting through the ink to the glass. Hold the glass close to the eye and look through the scratches at the slit held at arm's length. On each side of the slit there will be seen a series of spectra. Cover half of the slit with blue glass and the other half with red glass. There will now be seen a series of red images and also a series of blue ones, the red being farther apart than the blue.

The phenomenon exhibited in this experiment is called *Diffraction*. It is due to the interference of the rays which bend around opaque obstructions, the effects being made visible by the narrow spaces on the glass reducing greatly the intensity of the direct rays from the source of light.

---

[1] The light from ACB differs in phase half a wave-length from that reflected from DE, because the former is reflected in an optically dense medium next to a rare one, and the latter in an optically rare medium next to a dense one. This phase-difference is additional to the one described above.

XI. OPTICAL INSTRUMENTS.

560. The Human Eye is a nearly spherical ball of about nine-tenths of an inch in diameter, situated in a bony cavity called the *Orbit*. Fig. 292 represents a section of it from back to front, in which C is the cornea, P the pupil, I the iris, R the retina, etc. The outermost covering is a thick and horny substance, opaque, except in front. This opaque part is called the *Sclerotica*, and forms what is known as the *White of the Eye*. Its transparent portion, the *Cornea*,

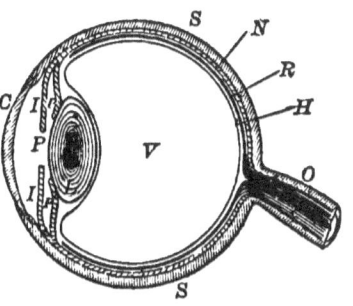

Fig. 292.

is more convex in shape, and appears as if set into the sclerotica, after the manner of a watch-glass in its frame. Behind the cornea is an annular and opaque diaphragm called the *Iris*, which constitutes the *Colored part of the eye*. The opening in this membrane is called the *Pupil*, the size of which varies with the intensity of the light. Between the iris and the cornea is a thin, transparent fluid, called the *Aqueous Humour*. Behind the pupil is a double-convex transparent body of uneven density, less convex on its anterior side, called the *Crystalline Lens*. The large cavity behind the iris is filled with the *Vitreous Humour*, a transparent, jelly-like substance, resembling the white of an egg. Lining the curved walls of this posterior chamber is the *Choroid Coat*, on which is spread the *Retina*, a membrane traversed by a network of nerves, the branches of the optic nerve. This choroid coat is saturated with a very black substance called the *Pigmentum Nigrum*, designed to darken the cavity of the eye, and to absorb all light reflected internally.

561. Sight Produced. — Rays of light diverging from the object enter the pupil and form an inverted image of the object on the retina, in the same manner as images are formed by small apertures (480). The two humours and the crystalline lens by their successive refractions give sharpness to the definition. The luminous waves incident on the retina give their energy to the nerve fibres, which transmit it to the brain, where in some mysterious way the sensation of sight is produced.

562. Perception of Color. — Microscopic examination shows that the retina is thickly studded with little projections called *Cones*, composed of minute cylindrical or rod-like structures of varying lengths. Each rod is connected with a nerve fibre. It is generally supposed that these rods are of three kinds, tuned to vibrate sympathetically with the three rates of ethereal undulations corresponding to red, green, and violet (544). By stimulating these rods in suitable proportions any color of the spectrum is seen.[1]

563. Color Blindness. — It was discovered by Dalton that many persons are unable to see certain colors. This defect is known as *Color Blindness* or *Daltonism*. It is undoubtedly due to some defect in the retinal structure, all the rods, the seat of color sensation, not being equally sensitive, or perhaps one or more classes of them being wholly inoperative. In most cases the defect is confined to inability to see red and green as such.

---

[1] "Sight," by Professor Joseph Le Conte, p. 61.

564. Distinct Vision. — The eye has the power of changing the curvature of the crystalline lens so that light from near as well as from distant objects may be focused on the retina. A normal eye is one which, in its passive condition, focuses parallel rays on the retina. If, because the eye-ball from front to back is too long, such rays focus in front of the retina, the eye is *Myopic*, giving rise to *Near-sightedness*. By intercepting the rays with a diverging lens of suitable power, the rays are focused on the retina and the vision is distinct. If, on the contrary, the eye-ball from front to back is too short, the focus of parallel rays will fall back of the retina, and the eye is *Hypermetropic*, one unable to see near objects, but able to see distant ones. This kind of an eye must not be confounded with a *Presbyopic* one, where through some functional defect the crystalline lens loses the power of adjusting itself to rays of considerable divergence. This defect is known as *Far-sightedness*, and is corrected by intercepting the light with a convex lens. Since a normal eye adapts itself without painful effort to the distinct vision of small objects, like the letters on this page, at a distance of about 25 cm. (10 inches), it is customary to speak of this distance as that of distinct vision, classifying eyes as *near-sighted* when the distance is less than 10 inches and *far-sighted* when greater.

565. Distance Estimated. — There are several circumstances which aid in forming judgments of distance. *First.* Experience has taught us to estimate distance by the amount of effort necessary to adjust the eye to secure distinct vision. Judgments based on this are tolerably accurate for distances comprised between five inches and twenty feet. Beyond this distance the changes in the effort to focus the eye are too small to be appreciable. *Second.* Experience has taught us to estimate distance by the amount of muscular effort necessary to direct both eyes on the object to secure single vision. It is reliable for distances not exceeding one thousand feet. *Third.* We estimate distance by the apparent size of the object, that is, by comparing the angle formed at the eye by lines drawn from the extremities of the object with that obtained in the case of a body of known size and distance. *Fourth.* Our estimate of distance is materially aided by noting the change of color and of brightness of objects caused by variations in the depth of air looked through.

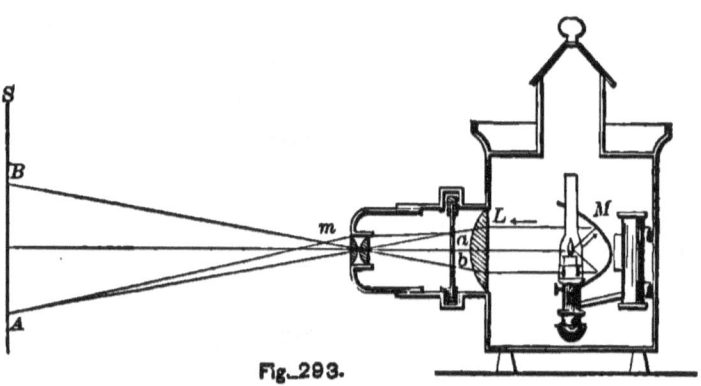

Fig. 293.

566. **Size Estimated.** — The size of an object is obtained by multiplying the size of the retinal image by the distance of the object from the eye in inches. Hence, an error in estimating distance produces a corresponding error in estimating its size. When there is no means of judging of distance, size cannot be estimated. This is illustrated by the varying estimates of the sun or moon by different observers. Indistinctness and great distance experience has taught us to associate together. Hence, at night,

we overestimate the distance of objects, and accordingly their apparent size is much exaggerated.

All conclusions regarding size and distance are but judgments, and hence may be greatly in error. The presence and the mode of distribution of surrounding objects often mislead the judgment and cause erroneous conclusions.

567. The Optical Lantern (Fig. 293) is an apparatus by which a greatly enlarged image of an object may be projected on a white screen. In its simplest form it consists of a box in which are placed a powerful light, a *Condenser L*, and an *Objective m*. The condenser is a single or compound lens having for its chief purpose the refraction of the light so as to bring as much as possible on the screen. The object, usually a drawing or photograph on glass, is placed near it and is thereby strongly illuminated. The objective is a simple or compound lens acting as a projecting lens in giving on the screen a real, inverted, and enlarged image of the object.

568. The Photographer's Camera (Fig. 294) is a box, adjustable in length, by means of the bellows *G*, the tube *DA* having an achromatic lens. If an object be placed in front of this lens, and the lens be properly focused, an inverted image of the object is formed on the plate *E* at the back of the box. By receiving this image on a plate coated with

Fig. 294.

a film containing the requisite chemicals, changes are produced in the coating which, by a process called "developing," becomes a permanent picture of the object on the plate.

Fig. 295.

569. The Compound Microscope (Fig. 295) is an instrument for seeing very small objects. In its simplest form, it consists of two convex lenses, the *Object-glass MN*, the one next to the object, and the *Eye-piece RS*, the one next to the eye (Fig. 296). The object-glass forms an enlarged real image of the object (where must the object be placed?), which is received by the eye-piece, so placed as to give a virtual image of this real image. (How must the eye-piece be placed with respect to this real image?)

570. The Astronomical Refracting Telescope is an

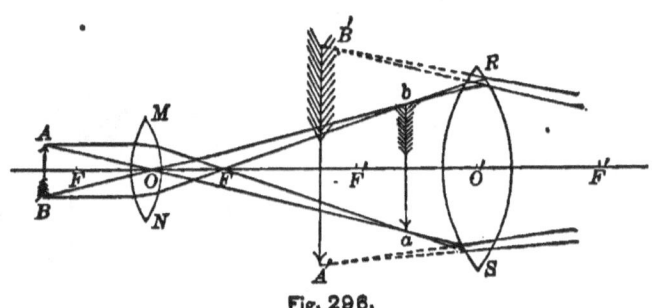

Fig. 296.

instrument for seeing distant objects. It is constructed on the same general plan as the compound microscope, except that the image of the object, formed by the objective, is smaller than the object, the magnification being confined to the eye-piece (Fig. 297). The office of the objective is to collect a large quantity of light in order that the real image formed may be bright, and thereby in a condition to sustain a large magnification without too great a loss of distinctness.

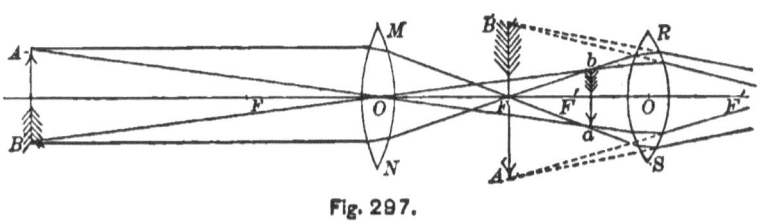

Fig. 297.

It will be seen from the figure that the object appears inverted, a matter of small moment in viewing celestial objects. In the common *Spy-glass*, the image is rendered erect by introducing near the eye-glass two convex lenses in such a way as to secure a second real image similar to the first, but reversed in position.

571. **The Opera Glass** is a double telescope, differing

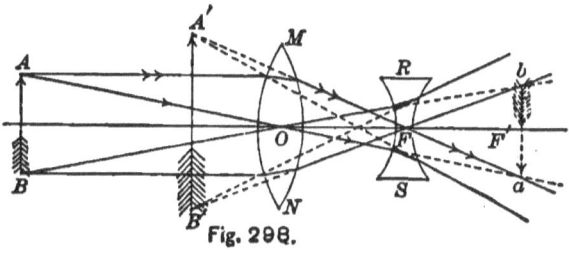

Fig. 298.

from the astronomical telescope in using a concave lens for an eye-piece. The rays issuing from the object are converged by the object-glass, and would form a real image beyond its principal focus, if they were not intercepted by the concave eye-piece (Fig. 298). By placing this eye-piece so that the rays incident upon it converge to a point beyond its principal focus, they will issue from it as diverging rays (526), and will consequently form an image (561) on the retina of the eye of an object apparently not as far distant as the real object.

DIAGRAMS OF METRIC MEASURE.

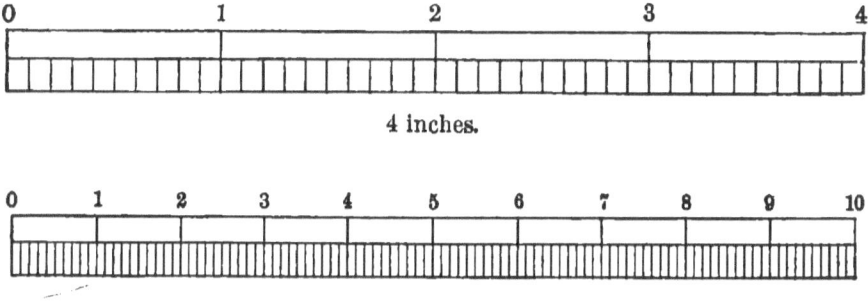

4 inches.

1 decimetre = 10 centimetres = 100 millimetres.

1 sq. decim. = 15.5 sq. in.
1 sq. in.   = 6.45 sq. cm.

A cube of water at 4° C., one of whose faces is this square, has a mass of one kilogramme. The volume is a litre. A cube of water at 4° C., with each of its faces one sq. cm., has a mass of one gramme. A cubic inch of water at 4° C. has a mass of 0.03611 lb. or 0.58 oz.

1 sq. in.

1 sq. cm.

TABLES FOR CONVERTING U. S. WEIGHTS AND MEASURES—CUSTOMARY TO METRIC.*

LINEAR.

	Inches to millimetres.	Feet to metres.	Yards to metres.	Miles to kilometres.
1 =	25.4000	0.304801	0.914402	1.60935
2 =	50.8001	0.609601	1.828804	3.21869
3 =	76.2001	0.914402	2.743205	4.82804
4 =	101.6002	1.219202	3.657607	6.43739
5 =	127.0002	1.524003	4.572009	8.04674
6 =	152.4003	1.828804	5.486411	9.65608
7 =	177.8003	2.133604	6.400813	11.26543
8 =	203.2004	2.438405	7.315215	12.87478
9 =	228.6004	2.743205	8.229616	14.48412

SQUARE.

	Square inches to square centimetres.	Square feet to square decimetres.	Square yards to square metres.	Acres to hectares.
1 =	6.452	9.290	0.836	0.4047
2 =	12.903	18.581	1.672	0.8094
3 =	19.355	27.871	2.508	1.2141
4 =	25.807	37.161	3.344	1.6187
5 =	32.258	46.452	4.181	2.0234
6 =	38.710	55.742	5.017	2.4281
7 =	45.161	65.032	5.853	2.8328
8 =	51.613	74.323	6.689	3.2375
9 =	58.065	83.613	7.525	3.6422

CUBIC.

	Cubic inches to cubic centimetres.	Cubic feet to cubic metres.	Cubic yards to cubic metres.	Bushels to hectolitres.
1 =	16.387	0.02832	0.765	0.35242
2 =	32.774	0.05663	1.529	0.70485
3 =	49.161	0.08495	2.294	1.05727
4 =	65.549	0.11327	3.058	1.40969
5 =	81.936	0.14158	3.823	1.76211
6 =	98.323	0.16990	4.587	2.11454
7 =	114.710	0.19822	5.352	2.46696
8 =	131.097	0.22654	6.116	2.81938
9 =	147.484	0.25485	6.881	3.17181

* Prepared by T. C. Mendenhall, late Superintendent of U. S. Coast and Geodetic Survey.

APPENDIX.

CUSTOMARY TO METRIC.—Continued.

CAPACITY.

	Fluid drams to millilitres or cubic centimetres.	Fluid ounces to millilitres.	Quarts to litres.	Gallons to litres.
1 =	3.70	29.57	0.94636	3.78544
2 =	7.39	59.15	1.89272	7.57088
3 =	11.09	88.72	2.83908	11.35632
4 =	14.79	118.30	3.78544	15.14176
5 =	18.48	147.87	4.73180	18.92720
6 =	22.18	177.44	5.67816	22.71264
7 =	25.88	207.02	6.62452	26.49808
8 =	29.57	236.59	7.57088	30.28352
9 =	33.28	266.16	8.51724	34.06896

WEIGHT.

	Grains to milligrammes.	Avoirdupois ounces to grammes.	Avoirdupois pounds to kilogrammes.	Troy ounces to grammes.
1 =	64.7989	28.3495	0.45359	31.10348
2 =	129.5978	56.6991	0.90719	62.20696
3 =	194.3968	85.0486	1.36078	93.31044
4 =	259.1957	113.3981	1.81437	124.41392
5 =	323.9946	141.7476	2.26796	155.51740
6 =	388.7935	170.0972	2.72156	186.62089
7 =	453.5924	198.4467	3.17515	217.72437
8 =	518.3914	226.7962	3.62874	248.82785
9 =	583.1903	255.1457	4.08233	279.93133

1 chain	=	20.1169	metres.
1 square mile	=	259	hectares.
1 fathom	=	1.829	metres.
1 nautical mile	=	1853.27	metres.
1 foot = 0.304801 metre,		9.4840158	log.
1 avoir. pound	=	453.5924277	grammes.
15432.35639 grains	=	1	kilogramme.

TABLES FOR CONVERTING U. S. WEIGHTS AND MEASURES—METRIC TO CUSTOMARY.

LINEAR.

	Metres to inches.	Metres to feet.	Metres to yards.	Kilometres to miles.
1 =	39.3700	3.28083	1.093611	0.62137
2 =	78.7400	6.56167	2.187222	1.24274
3 =	118.1100	9.84250	3.280833	1.86411
4 =	157.4800	13.12333	4.374444	2.48548
5 =	196.8500	16.40417	5.468056	3.10685
6 =	236.2200	19.68500	6.561667	3.72822
7 =	275.5900	22.96583	7.655278	4.34959
8 =	314.9600	26.24667	8.748889	4.97096
9 =	354.3300	29.52750	9.842500	5.59233

SQUARE.

	Square centimetres to square inches.	Square metres to square feet.	Square metres to square yards.	Hectares to acres.
1 =	0.1550	10.764	1.196	2.471
2 =	0.3100	21.528	2.392	4.942
3 =	0.4650	32.292	3.588	7.413
4 =	0.6200	43.055	4.784	9.884
5 =	0.7750	53.819	5.980	12.355
6 =	0.9300	64.583	7.176	14.826
7 =	1.0850	75.347	8.372	17.297
8 =	1.2400	86.111	9.568	19.768
9 =	1.3950	96.874	10.764	22.239

CUBIC.

	Cubic centimetres to cubic inches.	Cubic decimetres to cubic inches.	Cubic metres to cubic feet.	Cubic metres to cubic yards.
1 =	0.0610	61.023	35.314	1.308
2 =	0.1220	122.047	70.629	2.616
3 =	0.1831	183.070	105.943	3.924
4 =	0.2441	244.093	141.258	5.232
5 =	0.3051	305.117	176.572	6.540
6 =	0.3661	366.140	211.887	7.848
7 =	0.4272	427.163	247.201	9.156
8 =	0.4882	488.187	282.516	10.464
9 =	0.5492	549.210	317.830	11.771

APPENDIX.

METRIC TO CUSTOMARY.—Continued.

CAPACITY.

Millilitres or cubic centilitres to fluid drams.	Centilitres to fluid ounces.	Litres to quarts.	Dekalitres to gallons.	Hektolitres to bushels.
1 = 0.27	0.338	1.0567	2.6417	2.8375
2 = 0.54	0.676	2.1134	5.2834	5.6750
3 = 0.81	1.014	3.1700	7.9251	8.5125
4 = 1.08	1.352	4.2267	10.5668	11.3500
5 = 1.35	1.691	5.2834	13.2085	14.1875
6 = 1.62	2.029	6.3401	15.8502	17.0250
7 = 1.89	2.368	7.3968	18.4919	19.8625
8 = 2.16	2.706	8.4534	21.1336	22.7000
9 = 2.43	3.043	9.5101	23.7753	25.5375

WEIGHT.

Milligrammes to grains.	Kilogrammes to grains.	Hectogrammes (100 grammes) to ounces avoirdupois.	Kilogrammes to pounds avoirdupois.
1 = 0.01543	15432.36	3.5274	2.20462
2 = 0.03086	30864.71	7.0548	4.40924
3 = 0.04630	46297.07	10.5822	6.61386
4 = 0.06173	61729.43	14.1096	8.81849
5 = 0.07716	77161.78	17.6370	11.02311
6 = 0.09259	92594.14	21.1644	13.22773
7 = 0.10803	108026.49	24.6918	15.43235
8 = 0.12346	123458.85	28.2192	17.63697
9 = 0.13889	138891.21	31.7466	19.84159

WEIGHT.— *Continued.*

Quintals to pounds avoirdupois.	Milliers or tonnes to pounds avoirdupois.	Grammes to ounces Troy.
1 = 220.46	2204.6	0.03215
2 = 440.92	4409.2	0.06430
3 = 661.38	6613.8	0.09645
4 = 881.84	8818.4	0.12860
5 = 1102.30	11023.0	0.16075
6 = 1322.76	13227.6	0.19290
7 = 1543.22	15432.2	0.22505
8 = 1763.68	17636.8	0.25721
9 = 1984.14	19841.4	0.28936

FORCE.

[Wherever g, the acceleration of gravity, occurs in these relations, it is assumed equal to 980 centimetres or 32.15 feet per second per second.]

1 poundal = 13826.456 dynes = 0.031104 pound.

1 dyne = 7.2330×10^{-5} poundals = 0.225×10^{-5} pounds = 0.0010204 gram.

1 pound = 444520.58 dynes = 32.15 poundals.

1 gramme = 980 dynes.

1 grain = 63.503 dynes.

1 pound pressure per square foot = 478.49 dynes pressure per square centimetre.

WORK.

1 erg = 1 dyne-centimetre = 0.0010204 gramme-centimetre = 2.3730×10^{-6} foot-poundals = 0.7381×10^{-7} foot-pounds.

1 megalerg = 10^6 ergs = 2.3730 foot-poundals = 0.07381 foot-pounds.

1 kilogramme-metre = 7.2330 foot-pounds = 980×10^5 ergs.

1 foot-pound = 0.138255 kilogramme-metre = 1354.9×10^4 ergs = 32.15 foot-poundals.

1 foot-poundal = 421429.44 ergs = 0.031104 foot-pound.

ACTIVITY.

1 watt = 10^7 ergs per second = 44.2866 foot-pounds per minute = 0.10204 kilogramme-metre per second.

1 erg per second = 0.0000001 watt.

1 horse-power = 33000 foot-pounds per minute = 550 foot-pounds per second = 745.19 watts.

APPENDIX.

DENSITIES OF A FEW OF THE MORE IMPORTANT SUBSTANCES.

The following table gives the mass in grammes of 1 cubic centimetre of the substance:—

Substance	Density	Substance	Density
Agate	2.615	Hydrogen, at 0° C. and 76 cm. pressure	0.0000896
Air, at 0° C. and 76 cm. pressure	0.00129	Ice	0.917
Alcohol, ethyl, 90%, 20°C.	0.818	Iceland spar	2.723
Alcohol, methyl	0.814	India rubber	0.930
Alum, common	1.724	Iron, bar	7.788
Aluminium	2.670	Iron, cast	7.230
Antimony, cast	6.720	Iron, wrought	7.780
Beeswax	0.964	Ivory	1.820
Bismuth, cast	9.822	Lead, cast	11.360
Brass, cast	8.400	Magnesium	1.750
Carbon, gas	1.89	Marble	2.720
Carbon disulphide	1.293	Mercury, at 0° C.	13.596
Charcoal	1.6	Mercury, at 20° C.	13.558
Coal, anthracite	1.26 to 1.800	Milk	1.032
Coal, bituminous	1.27 to 1.423	Nitrogen, at 0° C. and 76 cm. pressure	0.001255
Copper, cast	8.83	Oil, olive	0.915
Cork	0.14 to 0.24	Oxygen, at 0° C., and 76 cm. pressure	0.00143
Diamond	3.530	Paraffin	0.824 to 0.940
Ebony	1.187	Platinum	21.531
Emery	3.900	Potassium	0.865
Ether	0.736	Silver, cast	10.424 to 10.511
Galena	7.580	Sodium	0.970
German silver	8.432	Steel	7.816
Glass, crown	2.520	Sulphur	2.033
Glass, flint	3.0 to 3.600	Sugar, cane	1.593
Glass, plate	2.760	Tin, cast	7.290
Glycerine	1.260	Water, at 0° C.	0.999
Gold	19.360	Water, at 20° C.	0.998
Granite	2.650	Water, sea	1.207
Graphite	2.500		
Gypsum, crys.	2.310		
Human body			7.000

INDEX.

Numbers refer to pages.

Aberration, chromatic, 351; lateral, 333; spherical, 327.
Absolute units of force, 34.
Absorption, 21; of heat, 138; spectra, 362.
Accelerated motion, 59.
Acceleration, 26; due to gravity, 61.
Achromatic lens, 352.
Action of points, 183, 196.
Adhesion, 13.
Air, state of, in sounding-tubes, 294.
Air-columns, laws for, 294.
Air-pump, 98; experiments with, 99.
Amalgamating zinc, 201.
Ammeter, 228.
Ampère, 224.
Amplitude, 65, 262.
Analysis of light, 350; spectrum, 364.
Antinodes, 292.
Aperture, effect on images, 309.
Arc light, 250.
Archimedes, principle of, 110.
Armature of magnet, 219; of dynamo, 246, 248.
Athermanous substance, 138.
Atmosphere, a unit of pressure, 95.
Atom, 1.
Attraction, law of, 50.

Attractive forces, 4.
Audibility, limits of, 285.
Aurora, 200; tube, 243.

Bacchus experiment, 99.
Balance, 72.
Barometer, 95; uses of, 96.
Batteries, connected in parallel, 226; connected in series, 225.
Battery, Bunsen, 209; chromic acid, 209; Daniell, 207; gravity, 208; Grenet, 210; Grove, 209; Leclanché, 211; plunge, 211; secondary or storage, 215; Smee, 207.
Beam of light, 308.
Beats, 282.
Bell, electric, 256.
Bladder-glass, 88.
Blake transmitter, 258.
Boiling-point, on thermometer, 122; of liquids, 150.
Boyle's law, 102.
Bright line spectra, 362.
Buoyant force of fluids, 111.

Calorie, 153.
Calorimetry, 153.
Camera, photographer's, 371.
Capacity, electrical, 184, 186; inductive, 181.
Capillarity, 19.

Carhart-Clark cell, 224.
Caustic, 328.
Cell, voltaic, 201.
Centre of curvature, 340; of gravity, inertia, mass, 51; of oscillation and of suspension, 67; optical, 340.
Centrifugal force, 57.
Centripetal force, 57.
Chemical changes, 6; effects of electric current, 212, 243.
Chemistry, 3.
Chord, major, 287; minor, 288.
Chromatic aberration, 351; scale, 288.
Chromic acid cell, 209.
Circuit, breaking and closing, 204; divided, 230.
Coefficient of elasticity, 12; of expansion, 142.
Coercive force, 167.
Cohesion, 13.
Coil, induction, 239; primary, 238; secondary, 238.
Cold by evaporation, 151.
Colloids, 24.
Color, 356; blindness, 368; complementary, 359; mixing, 358; not of the body, 357; of opaque bodies, 356; of transparent bodies, 357; perception of, 368.
Commutator, 247.
Compressibility of gases, 102.
Concave lens, focus of, 344.
Concave mirror, focus of, 322.
Concurring forces, 36.
Condensation, heat by, 152.
Condenser, 186, 242.
Condensing pump, 101.
Conduction of heat, 129; of electricity, 178, 222.

Conservation of energy, 48.
Consonant sounds, 299.
Constant force, 4.
Continuous force, 4.
Convection, 133.
Convex lenses, foci of, 342.
Convex mirrors, foci of, 324.
Cooling, Newton's law of, 138.
Critical angle, 337.
Crystallization, 14; expansion during, 143.
Crystalloids, 24.
Currents, electric, 200; chemical effects of, 212; extra, 238; heating effects of, 211; induced by currents, 237; induced by magnets, 235; laws of heat development by, 212; magnetic effects of, 216; mutual action of, 220; strength of, 224.
Curvilinear motion, 57.

Daltonism, 368.
Daniell cell, 207.
Declination, magnetic, 173.
Density, 113; optical, 331.
Deviation, angle of, 334.
Dialysis, 23; dialyser, 24.
Diamagnetic, 164.
Diathermanous substance, 138.
Diatonic scale, 286.
Dielectrics, 181.
Diffraction, 366.
Diffusion, 21.
Dipping-needle, 173.
Discord, 299.
Dispersion, 350.
Dissonance, 299.
Distance estimated, 369.
Distillation, 152.
Drum armature, 248.

INDEX. 385

Ductility, 16.
Dynamic unit of force, 34.
Dynamics, 25.
Dynamo, 244; compound, 248; series, 248; shunt, 248.
Dynamometer, 35.
Dyne, 34.

Earth, a magnet, 172.
Ebullition, 148.
Echo, 271.
Elasticity, 11.
Electric bell, 256; circuit, 204; current, 200; current detected, 203; light, 250; motor, 253; pressure, 224; spark, 185; telegraph, 253.
Electrical attraction, 174; capacity, 184; conduction, 178; distribution, 181; field, 180; machine, 189; measurements, 228; repulsion, 175.
Electricity, chemical effects of, 212, 243; heating effects of, 197, 243; luminous effects of, 243; magnetic effects of, 216; mechanical effects of, 197, 243; physiological effects of, 242.
Electrification, 175; by induction, 180; two kinds of, 176; nature of, 179.
Electrodes, 204.
Electrolysis, 212.
Electro-magnet, 218.
Electromotive force, 221, 224.
Electrophorus, 190.
Electroplating, 214.
Electroscope, 177; charged by induction, 181.
Electrotyping, 215.
Element, voltaic, 201.
Endosmose, 23.

Energy, 3, 46; indestructibility of, 48; kinetic, 46; measured, 47; potential, 47; transformations of, 48.
Engine, steam, 158.
Equilibrant force, 36.
Equilibrium, 52; kinds of, 53.
Erg, 44.
Ether, 306.
Evaporation, 148; disappearance of heat during, 151.
Exosmose, 23.
Expansion, coefficient of, 142; force of, 144; of gases, 141; of liquids, 141; of solids, 140.
Experiment, 5.
Extension, 6; its measurement, 7.
Eye, 367.

Falling bodies, 61.
Field, electrical, 180; magnetic, 169, 216.
Floating bodies, 112.
Flotation, 113.
Fluidity, 83.
Fluids, 2; equilibrium in, 92; perfect, 83.
Focal distance, measurement of, 349.
Focus, 321; of lenses, 342; of concave mirrors, 322; of convex mirrors, 324.
Foot, 7; foot-pound, 44.
Force, 4; buoyant, 111; classification of, 4; coercive, 167; composition of, 36; concurring, 36; graphic representation of, 35; how measured, 35; lines of, 169; molar, 4; molecular, 1, 4; parallelogram of, 37; resolution of, 39; units of, 34.

Fountain, 94, 108; vacuum, 100.
Franklin's experiment, 150.
Fraunhofer lines, 360; accounted for, 363.
Freezing mixture, 148.
Freezing-point, 122.
Friction, 71.
Fundamental tone, 292.
Fusion, latent heat of, 155; laws of, 146

Galvanometer, 228; tangent, 228; d'Arsonval, 229.
Galvanoscope, 204.
Gamut, musical, 286.
Gases, 2; as sonorous bodies, 294; compressibility of, 83; expansion of, 141; media for sound, 268; velocity of sound in, 278.
Gassiot's cascade, 244.
Geissler tubes, 244.
Graphic methods in sound, 302.
Gravitation, 50.
Gravity, 51; acceleration due to, 61; centre of, 52; cell, 208.
Grenet cell, 210.
Grove cell, 209.

Hardness, 16.
Harmonic motion, 262.
Harmonics, 293.
Harmony, 299.
Heat, absorption of, 138; capacity for, 154; conduction of, 129; convection of, 133; disappearance of during evaporation, 151; disappears during liquefaction, 146; disappears during solution, 147; distribution of, 128; due to chemical action, 128; due to collision, 126; due to compression, 127; due to electric current, 211, 243; due to friction, 127; effects of, 140; evolved during condensation, 152; evolved during solidification, 147; form of energy, 119; latent, 146; laws of radiant, 137; mechanical equivalent of, 159; radiant, 137; related to work, 157; sensible, 146; sources of, 126; specific, 154.
Heights, measured by boiling-point, 151; measured by barometer, 97.
Helix, 217; polarity of, 219.
Holtz machine, 191.
Horizontal line, 51.
Horse-power, 45.
Hydraulic press, 86.
Hydrometer, 115.
Hypothesis, 3.

Ice, latent heat of, 155.
Images, in curved mirrors, 324; in plane mirrors, 317; by lenses, 346; by small apertures, 309.
Impenetrability, 9.
Impulsive force, 4.
Incandescent lamp, 252; light, 250.
Inclination of magnetic needle, 173.
Inclined plane, 77.
Indestructibility, 11.
Index of refraction, 331.
Induced electric currents, 235; magnetism, 165.
Induction coil, 239; earth's, 172; electrically charged by, 180; electro-static, 180; self, 238.
Inductive capacity, 181.

INDEX. 387

Inertia, 9; centre of, 52.
Insulation, electrical, 179.
Intensity of illumination, 312; of sound, 273.
Interference of light, 364; of sound, 279; by reflection, 364.
Intervals, 286; of diatonic scale, 288.
Isoclinic lines, 173.
Isogonic lines, 174.

Joule's equivalent, 158.

Key-note, 286.
Kilogramme, 8; kilogramme-metre, 44.
Kilometre, 7.
Kinetic energy, 46; theory of heat, 119.

Lantern, optical, 371.
Latent heat, 146; of fusion, 155; of steam, 156; of vaporization, 156.
Law, physical, 3; Boyle's, 102; Ohm's, 221; of cooling, 138; Pascal's, 84.
Laws of motion, 32, 41.
Leclanché cell, 211.
Lenses, 339; achromatic, 352; foci of, 342; images by, 346; trace ray through, 341.
Lever, 71; mechanical advantage of, 73.
Leyden jar, 186; action of, 187; battery of, 189; seat of charge, 188.
Lichtenberg's figures, 196.
Light, 306; diffraction of, 366; interference of, 364; reflection of, 315; refraction of, 329.

Lightning, 198; rod, 199.
Lines of force, 169; direction of, 171.
Liquefaction, 145; disappearance of heat during, 146.
Liquids at rest, 92; compressibility of, 83; expansion of, 141; medium for sound, 268; velocity of sound in, 278.
Local action, 204.
Lodestone, 161.
Longitudinal vibrations, 261.
Loudness of sound, 273.
Luminous body, 306.

Machines, 70; efficiency of, 71.
Magdeburg hemispheres, 89.
Magnet, electro, 218.
Magnets, artificial, 162; action between, 163; action of magnets on, 165; compound, 164; electric currents induced by, 235.
Magnetic circuit, 219; dip, 173; effects of electric current, 216; field, 169, 216; permeability, 171; substance, 164; transparency, 164.
Magnetism, induced, 165; nature of, 166; terrestrial, 172; theory of, 168.
Major chord, 287.
Malleability, 15.
Manometric flames, 304.
Mass, 8; centre of, 51; its constancy, 11.
Matter, 1; properties of, 6.
Mechanical advantage, 70, 73, 75, 76, 77, 80.
Mechanics, 25.
Melting-point, 146.
Metric system, 7, 8.

388 INDEX.

Micrometer, 82.
Microphone, 257.
Microscope, compound, 372.
Minor chord, 288.
Mirrors, 317; plane, 317; spherical, 321.
Molar force, 4.
Molecule, 1, 2.
Momentum, 32.
Motion, 25; accelerated, 26; harmonic, 262; Newton's laws of, 32, 41; reflected, 42; uniform, 26.
Motor, electric, 253.
Multiple reflection, 319.

Needle, dipping, 173.
Newton's law of cooling, 138; laws of motion, 32, 41; rings, 365.
Nodes, 292.
Noise, 284.

Octave, 286.
Ohm, 222.
Ohm's law, 221, 224.
Opaque bodies, 307; color of, 356.
Opera-glass, 373.
Optical centre, 340.
Optical lantern, 371.
Organ-pipe, 294.
Oscillation, centre of, 67.
Osmose, 22.
Overtones, 293; laws of, 296.

Partial tones, 293.
Pascal's law, 84.
Pencil of light, 308.
Pendulum, 63; conical, 261; laws of, 65; seconds, 66; utility of, 68.

Permeability, magnetic, 171.
Phenomenon, 3.
Phonograph, 303.
Photographer's camera, 371.
Photometer, 313.
Photometry, 312.
Physics, 3.
Pigments, mixing, 359.
Pitch, 284.
Plumb-line, 51.
Polarity of magnets, 162.
Polarization, 205.
Poles, consequent, 163; distinguished, 163.
Potential, difference of, 185; fall of, 231; zero, 185.
Pound, 8, 34.
Poundal, 34.
Power, 45.
Pressure, atmospheric, 94; electric, 224; in fluids, 87; on bottom of vessel, 89; rules for computing, 91.
Principle of Archimedes, 110; of flotation, 113.
Prism, 333.
Pulley, 75; mechanical advantage of, 76; system of, 76.
Pump, air, 98; condensing, 101; force, 110; suction, 109.
Push-button, 256.

Quality of sounds, 297.

Radiant heat, laws of, 137.
Radiation of heat, 137.
Rainbow, 352.
Rays, 307.
Receiver, telegraph, 254.
Reflection, diffused, 316; multiple, 319; of light, 315; of sound, 270.

INDEX. 389

Refracted ray, to draw, 332.
Refraction, cause of, 334; index of, 331; of light, 329; of sound, 272.
Relay, 254.
Resistance coils, 230; electrical, 221; laws of, 222; of divided circuits, 231; unit of, 222.
Resonator, 276; Helmholtz's, 298.
Retentivity, 167.
Rods, vibration of, 300.

Scale, thermometric, 123.
Screw, 80.
Secondary or storage battery, 215.
Seconds pendulum, 66.
Self-induction, 238.
Sensible heat, 146.
Shadows, 308.
Shuttle armature, 246.
Sight produced, 368.
Siphon, 104; intermittent, 108.
Siren, 285.
Size estimated, 370.
Smee cell, 207.
Solenoid, 217; polarity of, 219.
Solidification, 146; heat evolved during, 147.
Solids, 2; density of, 114; expansion of, 140; media for sound, 269; velocity of sound in, 279.
Solution, heat disappears during, 147.
Sonometer, 290.
Sound, intensity of, 273; reflection of, 270; refraction of, 272; transmission of, 268; sources of, 266; velocity of, 277; waves, 269.
Speaking-tube, 274.
Specific gravity, 113, 116.

Specific heat, 154; determination of, 154; why substances differ in, 155.
Spectra, absorption, 362; continuous, 362; discontinuous, 362; laws of, 363.
Spectroscope, 361.
Spectrum analysis, 364; solar, 350; pure, 360.
Spherical aberration, 327.
Spy-glass, 373.
Stability, 55.
Standard cell, 224.
Statics, 25.
Steam, latent heat of, 156.
Steam-engine, 158.
Steelyard, 72.
Strength of electric current, 224; methods of varying, 225.
Strings, laws of, 291.
Substances, 1; classified, 2.
Surface tension of liquids, 16.
Sympathetic vibrations, 282.

Tangential force, 57.
Telegraph, 253; signals, 255; system, 255.
Telephone, acoustic, 269; Bell, 256.
Telescope, 372.
Temperament, 288.
Temperature, 119.
Tenacity, 15.
Theory, 4.
Thermal capacity, 154; unit, 153.
Thermometer, air, 125; limits of use, 125; location of fixed points, 122; mercurial, 121.
Thunder, 198.
Tone, fundamental, 292.
Torricelli's experiment, 94.
Torsional vibrations, 261.

Translucent bodies, 307.
Transmitter, Blake, 258.
Transparent bodies, 307.
Transverse vibrations, 261.

Undulatory theory of light, 306.
Unison, 299.

Vaporization, 148; laws of, 149; latent heat of, 156.
Velocities, composition of, 27; compounding, 28; graphic representation of, 27; resolution of, 30.
Velocity, 25; of light, 311; of sound, 277.
Ventilation, 135.
Ventral segment, 293.
Vertical line, 51.
Vibrating body, 260; effect of area, 275; locating, 261.
Vibrations, 65; classified, 260; complete, 260; forced, 282; modes of, 289; of bells, 302; of plates, 301; of rods, 300; of strings, 290; period of, 262; sympathetic, 282.

Vibration-rate measured, 285.
Viscosity, 83.
Vision, distinct, 369.
Volt, 224.
Voltaic cell, 201.
Voltameter, 224.
Voss machine, 195.

Water hammer, 63.
Water, latent heat of, 156; thermal capacity of, 154.
Wave-length, 263; of light, 356.
Waves, 262; combining, 279; graphic representation of, 262; on a medium, 264; through a medium, 264; velocity of, 266.
Wedge, 79.
Weight, apparent and true, 112; law of, 51.
Wheatstone's bridge, 232.
Wheel and Axle, 74; mechanical advantage of, 75; modifications of, 75.
Whispering-gallery, 271.
Wimshurst machine, 195.
Work, 44.

www.ingramcontent.com/pod-product-compliance
Lightning Source LLC
Chambersburg PA
CBHW030426300426
44112CB00009B/873